DISTRIBUTED GROUP COMMUNICATION
The AMIGO Information Model

ELLIS HORWOOD BOOKS IN INFORMATION TECHNOLOGY
General Editor: Dr JOHN M. M. PINKERTON, Principal, J & H Pinkerton Associates, Surrey (Consultants in Information Technology), and formerly Manager of Strategic Requirements, ICL

DISTRIBUTED
GROUP
COMMUNICATION
The AMIGO
Information Model

The AMIGO MHS+ Group

Editors:

HUGH SMITH, B.Sc., M.Sc.
JULIAN ONIONS, B.Sc., M.Phil.
STEVE BENFORD, B.Sc., Ph.D.

all of the Communications Research Group
Department of Computer Science
University of Nottingham

Sponsored by COST 11ter

ELLIS HORWOOD LIMITED
Publishers · Chichester

Halsted Press: a division of
JOHN WILEY & SONS
New York · Chichester · Brisbane · Toronto

First published in 1989 by
ELLIS HORWOOD LIMITED
Market Cross House, Cooper Street,
Chichester, West Sussex, PO19 1EB, England
The publisher's colophon is reproduced from James Gillison's drawing of the ancient Market Cross, Chichester.

Distributors:

Australia and New Zealand:
JACARANDA WILEY LIMITED
GPO Box 859, Brisbane, Queensland 4001, Australia

Canada:
JOHN WILEY & SONS CANADA LIMITED
22 Worcester Road, Rexdale, Ontario, Canada

Europe and Africa:
JOHN WILEY & SONS LIMITED
Baffins Lane, Chichester, West Sussex, England

North and South America and the rest of the world:
Halsted Press: a division of
JOHN WILEY & SONS
605 Third Avenue, New York, NY 10158, USA

South-East Asia
JOHN WILEY & SONS (SEA) PTE LIMITED
37 Jalan Pemimpin # 05–04
Block B, Union Industrial Building, Singapore 2057

Indian Subcontinent
WILEY EASTERN LIMITED
4835/24 Ansari Road
Daryaganj, New Delhi 110002, India

British Library Cataloguing in Publication Data
Distributed group communication: the AMIGO information model. —
(Ellis Horwood books in information technology)
1. Man. Communication with computer systems.
I. Title. II. Smith, Hugh. III. Onions, Julian.
IV. Benford, Steve. V. Series
004'.01'9

Library of Congress data available

ISBN 0–7458–0741–0 (Ellis Horwood Limited)
ISBN 0–470–21515–1 (Halsted Press)

Printed in Great Britain by Hartnolls, Bodmin

Contributors

Christian Huitema

INRIA
Centre de Sophia Antipolis
06565 - VALBONNE CEDEX
France

Yi-Zhi You

Dept Infa
EMSE
158 Cours Fauriels
42023 St-Etienne
France

Manfred Bogen,
Horst Santo,
Bernd Wagner,

Institut für Angewandte Informations–
Technologie (F3) der GMD Birlinghoven
P.O. Box 1240
D – 5205 St. Augustin 1
West Germany

Karl-Heinz Weiß

Forschungszentrum für Offene Kommunikations-
systeme (FOKUS) der GMD Berlin
Hardenbergplatz 2
D-1000 Berlin 12
West Germany

Jaime Delgado
Manuel Medina

ETSE Telecomunicacio
U.P. Catalunya
Apdo. 30.002
08080 - Barcelona
Spain

Andres G. Lanceros
Juan A. Saras Pazos

ETSI Telecomunicacion
Dpto. Ingeniena Telematica (DIT)
Ciudad Universitaria s/n
28040 - Madrid
Spain

Justo Carracedo Gallaroo
Miguel A. Nuñez

ETSI Telecomunicacion
Dpto. Ingeniena Telematica
Carr. Valencia Km. 7
18031 - Madrid
Spain

Jacob Palme

QZ
Box 27322
S-10254 Stockholm
Sweden

Steve Benford
Julian Onions,
Hugh Smith

Communications Research Group,
Computer Science Department,
Nottingham University, UK

Marko Bonac

Institute Jozef Stefan /E6
Jamova 39
61000 Ljubljana
Yugoslavia

Contents

List of Tables

List of Figures

Preface

Background to the AMIGO Project

The research described in this report was supported by the COST 11 ter action programme. COST (Coopération européene dans le domaine de la Recherche Scientifique et Technique) is a European framework for collaborative cross-border research which was initiated more than 15 years ago. It is supported by the 12 Member States of the European Community, and by cooperation agreements with the EFTA countries.

COST provides funds for groups of researchers, working at different institutions within Europe, to collaborate on a chosen research topic. The COST 11 ter action was concerned with teleinformatics research and it started in 1986. Nine projects were funded in total, including the AMIGO (Advanced Messaging in Groups) project. The AMIGO project was concerned with examining how support for collaborative working might be provided through the use of electronic distributed message systems. The project was split into three work areas entitled:

> Advanced Group Communication
> MHS+
> MMConf

The Advanced group was charged with the task of providing a conceptual framework for Group Communication activities. This work was seen to be inherently long term (i.e., not constrained by available communication capabilities and services).

The MHS+ group concerned itself with how group communication support could be obtained using present (or shortly available) OSI services.

The MMConf group examined 'on-line' group communication and the requirements for multi-media tools and support services.

This report describes the research activities undertaken by the MHS+ group.

Aims of the MHS+ research

Computer-based message systems have become almost commonplace in research organisations. Although they have increased the speed and convenience of the simpler forms of interpersonal communication, it has become clear that more sophisticated systems are required if the kinds of formal communication patterns used in offices and other user communities are to be fully supported. For example, whereas computer-based message systems (CBMS) support the exchange of single messages between one or more users, people think in terms of, and organise themselves to handle, 'streams' of communication over quite long periods of time.

With the imminent emergence of the X.400 message handling system and other standardized networking services, it has become important to investigate the requirements for such high-level group communication support in order to determine what kinds of future service enhancements need to be considered. The aims of the AMIGO MHS+ group were therefore to try and examine how group communication might be supported in a distributed environment by extensions to MHS and other services. The outcome of the research experience is presented in this report.

The collaborative work on the MHS+ project itself demonstrated the need for, and use of, better communication support facilities. Members of the group utilized CBMS to collaborate via interpersonal mail and distribution lists. This experience led to insights into what kinds of support would be required for different applications. Furthermore, parts of the research model described in this report were prototyped via a set of pilot experiments involving some of the MHS+ institutions.

The structure of the Report

This report describes an abstract model for group communication support and an outline implementation architecture. The report consists of a set of chapters, each originally authored by different members of the group. However, the final contents of the chapters often changed as a result of group discussion. Therefore, where appropriate, contributors as well as authors are indicated at the beginning of each chapter.

The structure of the report is best understood with reference to the diagram on the opposite page.

The different *User Requirements* which need to be supported within electronic communication systems are described in chapter 1. These requirements have been examined both by the AMIGO MHS+ group and by the AMIGO Advanced group. The latter has produced a *Group Activity Model* which provides a general means for the modeling of group communication processes.

The study of user requirements identified the need to produce an abstract model of group communication support in order to divorce implementation issues from conceptual issues.

Chapter two describes aspects of this model, the *Conceptual Data Model* and the need for a *Global Naming Scheme*. The Conceptual Data Model

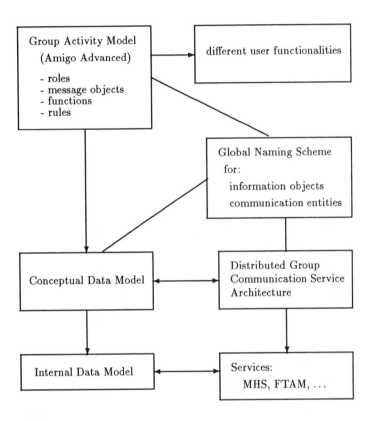

describes information relevant to communication in terms of *Information Objects* and takes a global view of information which is independent of specific user applications or the support system (e.g., storage and distribution) implementation. Communication Entities (e.g., individuals, groups, agents) access Information Objects during communication and both objects and entities are required to be identified by globally unique *names*.

Chapter 3 discusses a distributed *Group Communication Service Architecture* that is based on the conceptual data model.

At an implementation level, information objects must in principle be globally accessible (i.e., via object stores and replication schemes). The exact *Internal Data Model* will be dictated by the functionality of the available storage and distribution services. Chapters 4 to 6 discuss the nature of the internal data model and the utilisation of *ISO services*.

Chapter 7 discusses the experiences of members of the AMIGO MHS+ group in producing prototype (pilot) versions of parts of the Group Communication Service Architecture.

Chapter 8 briefly outlines the kinds of advanced group communication activities that are not addressed in the architectural chapters.

Notes for the Reader

This report assumes some familiarity with terms used in computer-based messaging systems and certain OSI standards. If you are unfamiliar with these areas, you should read chapter one and then appendix A. The appendix is intended to supply the necessary background and introduce its terminology.

The report consists of a series of papers which have a common theme. However, they should primarily be treated as a source of *ideas* and not as a fully worked out solution. This is because of the necessarily limited nature of the AMIGO research collaboration. The MHS+ group met approximately ten times during a two year research period and corresponded between meetings through the medium of electronic mail. This kind of collaboration is very fruitful in generating and interchanging ideas between researchers in different laboratories. However, the nature of the collaboration meant that not all the ideas that emerged could be pursued to their logical conclusion. So it is with the material presented in this report.

Acknowledgements

Overall direction to the AMIGO project was provided by Horst Santo. Jacob Palme was the chairman of the MHS+ group.

The final editing of this version of the report was performed by Steve Benford, Miguel Nuñez, Julian Onions, Hugh Smith and Karl Heinz Weiss. The production of this report was undertaken at Nottingham University by Steve Benford, Julian Onions and Hugh Smith.

Hugh Smith
December 1988

Chapter 1

An Introduction to Group Communication Service Requirements

Author: Hugh Smith

Contributors: Karl-Heinz Weiss

               Manfred Bogen

1.1 Introduction

The last few years have seen a dramatic growth in the use of electronic communication systems for information transfer. Although telex still has the greatest "reach" it has been technically superceded by the fax and the computer-based message service. In the modern office the computer-based message service (CBMS) is particularly important since information is transmitted in a revisable, processable form. Thus the contents of a message may be processed and/or re-used for other purposes (e.g., editing collected contributions, forwarding inquiries). This capability becomes particularly significant when several individuals utilize CBMS to engage in various types of co-operative work effort the dominant pattern of organizational activity. However, the effective support of organizational activities requires far more than just the availability of a message transfer system. It requires a set of information processing tools and services that facilitate the many different kinds of structured communication activity (e.g., document circulation, conferencing, scheduling) that are undertaken by work groups.

The focus of the AMIGO MHS research is to define such group-based communication services. We use the term *group communication* to imply any user-centred message-based activity that takes place over a period of time, involves several individuals, and models user requirements based on organizational or social rules. The goal of the AMIGO MHS+ work has been to describe a number of typical group communication activities and

define the necessary tools and services for their support in a distributed
CBMS environment. The assumed CBMS environment is the distributed
inter-personal message handling system defined by X.400/MOTIS [MHS84c,
MHS87a]. The CCITT/ISO defined Electronic mail system which defines
the exchange of units of information between users or their agents in a
store-and-forward system. However, this choice does not preclude the consideration of more interactive group communication support services based
on connection-oriented modes of communication.

In order to try to justify the need for more comprehensive group communication services than those currently available we will examine briefly some
existing CBMS applications.

1.1.1   Examples of existing structured communication appli-
cations

At present a number of electronic message systems provide the necessary
low-level distributed support for a variety of higher-level structured group
communication activities. Examples of such activities within the academic
community are:

- news distribution (e.g. USENETmoderated digests)

- conferencing (e.g. Notes)

- information storage and remote retrieval (e.g. The CSNET style info
  server)

However, there is no direct high-level processing support for these activities within the messaging system. Consequently, functionality is limited
and automatic error detection and correction is, at best, poor. In certain
application areas (e.g., conferencing) there are dedicated systems which do
support group communication but they often do not interwork well with
other systems to provide a *distributed* communication service.

In fact the shortcomings of existing systems can be described under the
following headings:

- limited functionality

- lack of supporting services

- standardization and inter-operability

- centralized versus decentralized resources

- user adaptability and control

Limited functionality

It is a fundamentally desirable requirement that communicated information should be able to be used for more than one function. Thus the output from one activity should be usable as the input to another (e.g. a reply to a keyword search request being re-used to order a copy of the retrieved document). Existing systems are frequently prevented by both technical and functional incompatibilities from achieving this kind of economy. For example, to achieve re-usability of information in a message, user must often transform/copy blocks of text from one message to another.

Lack of supporting services

Many kinds of services are required to support structured communication activities in addition to the basic information transfer capability provided by a message delivery system. Examples include services for the management, organisation and identification of sets of messages/activities; distribution control; authentication; storage, retrieval and many others. The introduction of such services will require messages to carry more structured control information than at present e.g. references to body parts of preceding messages (or parts of the bodies).

Standardization and inter-operability

As yet there are no standardized services to support group communication. Although "islands" of communication exist within many networked communities, little interworking is possible for other than basic messaging functions. As a result of this limitation on inter-operability, systems often fail to achieve the critical level of usage which would enable electronic methods of communication to be fully exploited. Although the adoption of basic communication standards such as X.400 and X.500[DS87] (the associated directory service) provide basic building blocks, higher layer tools are required.

Centralized versus decentralized resources

As mentioned above, there are a few examples of successful structured communication systems (e.g. conferencing – COM, EIES) but these often have a centralized resource architecture. In the future the sheer size and political/geographical separation of networked communities will demand that the resources and the management of group communication activities be distributed. However, the replication and co-ordination of information at distributed sites is a formidable problem – not least, because one cannot guarantee a homogeneous user agent environment (i.e. different users will often have different interface facilities).

User adaptability and control

Most existing electronic message systems are inflexible and not easy for the end-users to adapt to their specific purposes. It is important to achieve a level of adaptability that can support not only a generic activity but also the individual and his/her specific environment.

1.2   Definition Group Communication Activities

So far we have talked in fairly general terms about the need for group communication support without attempting to define it precisely. Several researchers have tried to formalize mass communication requirements (e.g, see[PS87, MGT⁺87]) including recently the AMIGO Advanced group[Ami89] and the COSMOS project (e.g. [Wil87]). Elements of some theories have been incorporated into research systems and at least one commercial system Co-ordinator[WF86].

However, given the limited size and duration of the AMIGO project the approach adopted by the MHS+ group is more short term and relatively pragmatic. A small number of existing communication scenarios were examined in order to abstract general support requirements that need to be considered in a network based message system context. From these scenarios it was possible to specify what additional service protocols need to be provided to support group communication in the X.400 environment.

Two general kinds of question were formulated:

- What kinds of group communication activities have been (and might be) attempted using electronic message systems and with what degree of success?

- What are the basic elements and component functions employed in performing these activities?

and more specifically from an implementation viewpoint:

- How may these activities be realised given the available standard services?

The following sections examine these questions.

1.2.1   Examples of Group Communication Activities

The following are examples of different kinds of group communication activities that might be based around message exchange systems:

- News/Notice Distribution (e.g. information distribution, journals)

- Bulletin Board (e.g. expert advice, for sale/want)

- Conference (e.g. discussion or decision group)

- Decision Taking (e.g. voting)

- Project Planning

- Scheduling (e.g. date of meeting booking)

- Co-ordinated Work (e.g. joint editing)

There is an increasing level of complexity here – a system which supports conferencing should be able to support notice distribution (but the reverse requirement is not true). We need to establish the functional support requirements for these kinds of communication activities. In this chapter we will briefly consider the first three of these activities.

News/Notice Distribution

This implies a distribution facility which allows a group of subscribers to receive information on topics for which they have registered an interest. (An office example would be a work group circulation list.) The "news" typically consists of discrete units of information circulated either, one per message, or in a "digestified" form in which many items are encapsulated within one envelope. In either case a moderator could undertake to receive contributions, edit them (on the basis of interest/relevance) and forward them to the subscribers. Both kinds of distribution format need an agent to act a distribution mechanism information resource and ensure that errors are handled correctly a distribution manager. The subscribers are largely passive (i.e., they just read the news articles). In summary the facilities required are shown in Table 1.1.

Facilities	Example of use
Set-up Facilities	creation, deletion of news distribution
Membership control	registrations, updates, deletions
Submission policy	who does or can/cannot put material in news
Distribution policy	administration, error control
Information service	advertises news availability, handle enquiries
News topic control	topic definition, maintenance
Archive service	back issues service

Table 1.1: Distribution Facilities

Bulletin Board

This is a slight variant on the above in which additional services are provided to handle a more participatory style of communication. Here the recipients get the chance to reply to/comment on specific items passed around on the distribution. An office example might be the communication related to setting up the location and time of a meeting. A more complex example of the kinds of facility that may be required is the "classified adverts" section of most small papers/magazines. Here, many subscribers engage in dialogue about specific items. The items do not have a long life or a typically a complicated topic structure. The *additional* support facilities required are shown in Table 1.2.

Facilities	Example of use
Follow-up service	reply to referenced item
	distribution of replies to all subscribers
Topic management	facilities for deleting/closing off finished topics

Table 1.2: Additional Facilities for Bulletin Board

Conference service

Another increment on the facilities introduced above. In a typical conference system the communication has more of a discussion and review flavour. The *process* of reaching an agreed position may be as important as the position reached. An example might be the process of producing a technical specification of a product. The records of the discussion are maintained so that they are always available to each participant. More comprehensive topic management facilities are available than with the two previous examples so that any "streams" of conversation are kept clearly separated and, if necessary, new side conferences encompassing these and sub-groups of the participants can be initiated.

Facilities	Example of use
Membership Information	who reads what
Conference Security	open/closed/restricted conferences
	visible, invisible
Conference Topic control	structuring of dialogue into 'conversations'
	attachment of keywords to dialogue units
	reorganization of topics - new conference
Selection control	filter out topics - per user
Allied operations	voting, joint editing

Table 1.3: Examples of Facilities and Use

1.2.2   Elements

In the activity scenarios described above we can identify at least two kinds of elements in the communication description: Entities and Functions. An

Entity can be defined either as a communication agent (i.e. user or machine), information object (e.g. message, document) or resource (e.g. store). Functions such as reading, copying and storing are performed as part of the communication activity that entities engage in[1].

Entities

The three kinds of entity identified above are described as follows:

Agent Agents are often the end-points of the communication activity (i.e. the sender/receiver) but can also be involved in other phases (e.g. the distribution manager). An important concept is that of a *Role*. A Role agent may be regarded as an alias entity which may be bound, temporarily or permanently, to one or more people. Thus there may be secretary or manager roles which are identified with different people at different times. A 'group' could be defined as the set of role agents associated with a particular activity.

Information Objects Conventional message systems are designed to handle individual message. However, the communication activities identified in the previous section focus on more complex structures (e.g. streams of correspondence, documents, news articles). The *effective* support of group communication requires that such composite structures and their component parts be uniquely referenceable. Thus it should be possible to access or refer to all of the following:

```
message\index{message} 452
all replies to message\index{message} 452
all message\index{message} to do with bug fixing
the current set of ODA\index{ODA} standards
```

In a conference system it is likely that the set of objects so referenced would change from access to access. As can be seen in the last two examples, this would be a very useful capability.

Resources Certain kinds of resources will be associated with agents and information objects. For example, an *archival store* for keeping transactions or back issues of articles. Other tools might provide processing resources – e.g., filters, authenticators. These resources will usually be associated with performance of the Functions identified below.

Functions

There are many communication related functions that may be performed by Entities as part of a group communication activity. A partial list is shown in Table 1.4.

---

[1]The AMIGO Advanced research group have examined many of the issues related to what is reported here in greater depth [Ami89].

reading	categorizing	filtering
copying	annotating	editing
composing	replying	sending
distributing	searching	referencing
collating	assembling	storing

Table 1.4: List of Group Communication Functions

Ideally, these functions should be applicable to both the simple and composite forms of the information objects identified above.

However, we have not chosen to distinguish *where* or *how* these activities are performed. The important point is that the *same* functions are often carried out by different communication roles (either at the same or different times) in *different* environments. For example, assuming news distribution in a public network the functions shown in Table 1.5 might be performed by several entities.

OPERATIONS	Locus of Effect (environment)		
	Public	Group	Individual
	(e.g. writer)	(e.g. editor)	(e.g. Reader)
Categorizing	x	x	x
Filtering	x	x	x
Distributing	x	x	
Retrieving	x	x	x
Comparing	x	x	x
Transforming	x	x	x

Table 1.5: Functions for news Distribution

Thus although it may not be appropriate to use the same mechanisms for each function, given a distributed resource environment, there would be obvious benefits if the operation was conceptually identical in each environment. Figure 1.1 shows the sequence of functions that an *editor* and the *reader* of a given news distribution might invoke those processes.

If the definition of these functions was standardized then considerable gains of efficiency might follow from the ability to utilize similar (or even the same) mechanisms in different environments.

We will now turn to the architectural issues surrounding the implementation of our group communication activity examples.

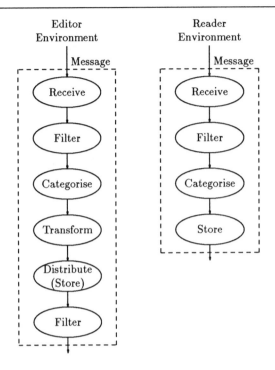

Figure 1.1: User Environments

1.3 Distributed Architectures for Group Communication

Section 1.2.1 discussed several examples of group communication activities and the previous section outlined some of the abstract elements that can be used to describe such activities. This section briefly discusses some of the architectural requirements for group communication.

One of the principle assumptions underlying the work described in this report is that group communication needs to be supported in distributed environments. To cater for the thousands of potential users and achieve the most widespread use requires that group communication facilities must be *based* on emerging Open System Interconnection (OSI) services. At the present time there are relatively few such services but this is expected to change quickly in the near future.

The AMIGO MHS+ group examined the four services listed in Table 1.6 with a view to employing them in a group communication service architecture. Of these four, two have been progressed to the status of full ISO standards (MOTIS and FTAM), and two (DS and DFR) are yet to be fully defined. These services are referenced throughout this report and it is assumed that the reader has some passing familiarity with them. However,

Service	Useful for
X.400 MHS	Message transfer, some distribution list expansion
X.500 DS	Global access to entity names, information links
FTAM	File Transfer Access and Management
DFR	Document Filing and Retrieval

Table 1.6: Supporting services

for those who are unfamiliar, a summary of the services is given in appendix A. It should be noted in passing that all of these services are the subject of much debate and revisions are constantly being planned. Therefore reference should be made to the relevant standard documents themselves for accurate information.

An examination of these ISO services reveals that they are not sufficiently functional to enable the kinds of group communication activities identified previously to be performed in other than an ad hoc way. For example, communication among groups (n to m) is not supported in MHS apart from a rather simple distribution list expansion mechanism. Furthermore, there is no concept of a sequence of group related activities that underpin individual acts of communication such as message sending. In summary, limitations of the MHS include:

- No concepts for handling roles, rules, memberships, topics, targets and personal working methods.

- Complex working procedures like appointment planning or office procedures are not supported.

- Communication contexts are only weakly supported (CrossReferences, InReplyTo, Obsoletes etc.).

- The information objects in MHS are bound to messages, probes or reports.

The limitations also extend to preventing the MHS from efficiently interworking with other services. For example, the distribution of messages within a large group of geographically separated members (e.g. in different countries) is usually realized by nested distribution lists. This allows optimization of message transfer at the cost of storage inefficiency (because all messages are stored by each group member).

This limitation could be overcome by the availability of multi-user long term message archives (perhaps using FTAM or DFR). This would allow the optimization of group related message storage. However, in contrast to the distribution list solution, each member of the group would need to explicitly request **access** to the archive to read messages. Thus, on the one hand, there are distribution lists with optimized communication costs and increased storage costs. On the other hand, there are long term archives for a group with decreased storage costs and increased costs for the communication with the store (i.e. requests and results).

In principle, a user of a group communication system should be able to utilise either mechanism in order to satisfy his/her special needs - e.g. depending on location, access to host machine, daily work.

However, it is more difficult to envisage how the co-ordination of of co-operative activities within groups can be supported. This requires another layer of functionality above that defined by existing ISO services. For example, the following additional facilities need to be provided:

Information Object to Message Mapping
Directory Services for describing and managing Groups
Group Distribution Service
Authentication and Access Control mechanisms
Activity Co-ordination Managers

The rest of this report is concerned with the subject of how some of these facilities might be provided.

1.4 Summary

This chapter has outlined the kinds of group activities that need supporting in future communication systems. Subsequent chapters extend this discussion and propose mechanisms for supporting such activities.

The following chapter discusses an abstract data model for describing group communication. This is followed by chapters that discuss possible architectural implementations of the data model.

Chapter 2

The Global Information Space

Author: Steve Benford

Contributors: Christian Huitema
                Horst Santo

2.1   A menagerie of entities and objects

This chapter specifies an *information model* for group communication. The model provides a tool for describing the information pertinent to communication and can be used to model a wide variety of group communication processes. Examples of this information might be messages, documents or conversations.

The AMIGO MHS+ view of communication is information oriented and centres on the access of *Information Objects* by *Communication Entities*. This means that communication is achieved by the coordinated, shared access and manipulation of sets of information by groups of entities. It is important to understand both the role and structure of Communication Entities and Information Objects.

A Communication Entity is an entity which takes part in a communication process. Various types of Communication Entity appear in the context of group communication. For example: users, groups, roles and the agents necessary to provide a Group Communication Service.

An Information Object is a unit of information which is exchanged or manipulated during a communication process. Many types of Information Object may appear within the context of group communication. For example: messages, documents and conversations.

The structure and naming of Communication Entities and Information Ob-

jects is clearly of great importance. The structure and naming of information representing Communication Entities is already considered by the
X.500 standard for Directory services [DS87]. The goal of this chapter is to
specify an information model describing the structure and naming of Information Objects. To this end, the remainder of this chapter is structured in
the following way:

- Section 2.2 considers the naming of both Communication Entities and
  Information Objects.

- Section 2.3 describes the concepts of environments which provide logical information spaces for group working.

- Section 2.4 outlines the data model describing the abstract structure
  of a wide variety of information objects.

- Section 2.5 considers the structure of atomic objects representing the
  basic building blocks of the data model.

- Section 2.6 and section 2.7 describe compound objects representing
  sets of information objects.

- Section 2.8 briefly considers the issue of access control applied to the
  data model.

2.2   Naming

This section examines the critical issue of naming applied to both Communication Entities and Information Objects.

2.2.1   Naming of Communication Entities

One of the roles of the X.500 directory service is the naming of Communication Entities. The directory will assign globally unique and unambiguous
distinguished names to entities such as persons and groups and it is therefore
natural to adopt these distinguished names when a unique reference to these
entities is needed. Thus, Communication Entities will be named by X.500
directory names within the AMIGO MHS+ model.

2.2.2   Naming of Information objects

There is no theoretical limit on the classes of entity which can be represented
in the X.500 directory service and, therefore, one could devise distinguished
names for information objects. However, this would mean registering all
information objects with the directory, which could be highly expensive
(imagine registering all IP-messages). In general, the directory has not been

designed for efficient registration of short lived Information Objects that are exchanged within groups (e.g. IP-messages). However, one could register a few important Information Objects in the directory, so that a wide audience could obtain their descriptions and locations.

Information Objects are the units of information that will be exchanged within groups and have to be uniquely identified in order to permit their potential designation for cross references, retrieval from a storage system or shared access by a number of services. Information Objects may have different identifiers in different application contexts. For example, a document may have an ODA document id, but could also be accessed via a file name from an FTAM filestore and be registered in the directory with a distinguished name. Messages will be given IPM-ID within the X.400 P2 protocol. Thus, we propose that the names of Information Objects within the AMIGO MHS+ model will be structured as two components: a context identification and a context specific identifier as shown below.

```
IO name = <context identification><context specific identifier>
```

The context identification can be a directory distinguished name. For example, the context identification for IP-Messages might be derived from the OR-Name component of the IPM-ID and the context ID for documents might be derived from the FTAM file store or the ODA application.

The context specific identifier will be the *.local* component of the IPM-ID, the FTAM file-name or the ODA document identifier. It can be omitted from the distinguished name as registered in the directory.

2.3   Environments

Previous sections have outlined the role of Communication Entities and Information Objects within group communication. The AMIGO MHS+ view is that communication occurs due to some interaction between them. There may be many different modes of interaction describing different flavours of communication (e.g. conferencing, bulletin board and distribution list) and the model developed in this work should be general enough to support all of these.

Interaction between Communication Entities and Information Objects occurs within an *environment* which specifies certain properties of the interaction. An environment is the binding of a group of users to a collection of Information Objects, and to the roles or protocols adequate for accessing these Information Objects.

In addition, an environment binds conceptual entities and objects with the application entities actually implementing the communication service. For example, *Message Transfer Agents.* However, new sorts of application entity may be introduced in the future. The environment therefore provides the glue which binds the conceptual model dealing with abstract entities and objects to the internal model dealing with the agents and services which provide communication.

Thus an environment may specify groups of Communication Entities, sets of Information Objects, access policies, management policies and the names of application entities responsible for the specified entities and objects. Environments form the cornerstone of the AMIGO MHS+ model because they bind together the abstract data model (section 2.4) the architecture (chapter 3) and the internal model.

An environment has a description which contains a set of attributes made publically available by registration in the directory service. The use of the directory service for holding environment descriptions implies that environments are named by X.500 directory names.

This list of attributes might include:

- A textual description of the environment,

- A description of the Communication Entities forming groups within this environment.

- Access control rules specifying general access permissions to information associated with the environment.

- The names of Storage System Agents, Distribution Lists and Group Communication System Agents relevant to communication within the environment.

- The names of notable Information Objects within the environment (e.g. notable documents)

- Information about the organization of the information within the environment, e.g. the types of attributes that can be used for searching Information Objects.

An environment is associated with a set of Information Objects some of which could be used as a basis for modelling the equivalent of *computer conferences* or *bulletin boards.* The description of the conference could be provided within the environment along with the distribution policy (e.g. a set of distribution list through which new entries in the environment must be distributed through the members). The contents of each conference would be represented by a set of Information Objects. For example, communication within the AMIGO MHS+ group might be represented by an environment describing the membership of the group, various agents associated with storage, coordination and distribution and a number of notable Information Objects representing different conferences within the group. All of this information would be globally visible via the environment representation in the directory service.

2.4   A Data Model for Group Communication

2.4.1   The role of the data model

Communication involves the **sharing of information**. Information might be represented as messages transferred between groups of humans or as data in some storage system which can be accessed by many people. These two views are different sides of the same coin. The fact is that if many services are to act together to support group communication they must be able to manipulate shared communication information in some way.

The data model has two main purposes:

- Describing the structure of information in an abstract and global manner.

- Specifying abstract operations to manipluate this information.

The data model therefore provides an abstract modelling tool for information objects. Most people have a personal understanding of what objects such as messages, documents and conferences might be. These objects are the basic building blocks of group communication. It is therefore necessary to define them precisely. Furthermore, users will wish to refer to sets of these basic objects. Examples are:

- Sets of messages grouped by recipient or sender.

- Sets of documents with the same author or subject

- Sets of related objects such as all messages in reply to another message or all messages obsoleting another message.

Some of these sets may be defined in a well known way such as *conversations, bulletin boards* or *conferences*; others may be defined when searching for information and have a short life time. The data model must specify a method for defining and manipulating sets of objects in a general way. Thus, the treatment of sets of related information is an important part of the design of the model.

2.4.2   Layered view of the AMIGO data model

The design of the data model¡ for the AMIGO group communication environment can be divided into several layers as shown in Figure 2.1. This model allows the global view of data called the *conceptual schema* to be considered independent of the implementation view called the *internal schema*.

The conceptual schema defines the data model in abstract terms which are independent of the specific implementations of storage services. It is concerned with identifying the objects which are shared in the AMIGO environment and the operations which manipulate them without defining how they might be represented in existing AMIGO services. The conceptual schema presents a global view of data without considering issues concerned with its distribution.

The internal schema specifies the methods of representing the conceptual schema in the AMIGO distributed environment. It is concerned with issues such as locating objects, replicating objects, distributed update and consistency. The internal schema must consider the properties of services which will implement the distributed storage of communication information.

The figure indicates that the conceptual schema supports many *user views* of information. A user's view is a portion of the conceptual schema. The figure also shows that the internal schema is mapped to *storage schemas* of specific service implementations. These mappings specify the details of implementing the AMIGO storage model using specific underlying services.

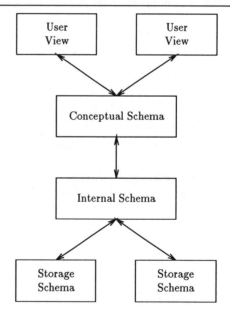

Figure 2.1: Layered view of Amigo information

This chapter specifies the conceptual schema and is not concerned with issues belonging to the internal schema or to specific user views of data. The conceptual schema must be defined with enough functionality to support the AMIGO framework and the following section describes some of the requirements which must be satisfied by the conceptual schema.

2.4.3   Requirements of the data model

The following paragraphs identify some of the requirements which must be met by the AMIGO conceptual data model.

- The data model must represent basic communication objects such as *messages* and *documents.*

- The data model must represent well known sets of objects such as *conferences* and *conversations.*

- The model should allow the expression of more general sets based on the structure of and relationships between objects. Examples of these sets are :-

  ```
  documents with fred as the author
  ```

  and

  ```
  messages sent by fred obsoleting messages sent by joe
  ```

  The mechanism for expressing sets should be as general as possible.

- The model should allow users to search for information matching criteria based on the set structure above. For example, search for

  ```
  conferences containing messages created after July 1987
  ```

  (this might be used to locate all conferences which contain unread messages).

- Objects should inhabit a namespace of globally unique names which can be used by programs and users accessing information.

- Objects should exist within environments and should be maintained and accessed according to policies associated with these environments.

- The model should support access control to Information Objects. Access controls should determine which information is globally visible thus providing *privacy* and should specify who can manipulate information thus aiding *integrity.*

- The model should support the functionality to read, search for, create, delete and modify objects, sets of objects and access controls.

2.5   Atomic objects

This section specifies the structure of *atomic objects* and the operations
used to read, create, delete and maintain them. Atomic objects are the
basic building blocks of the data model and are used in creating *compound
objects* or *sets* as will be described later.

Each atomic object represents a single real world object such as a *document*
or *message*. The essence of these objects is that although they may have
quite complex structures (e.g. the structure of an X.400 IP-Message) they
are viewed as a single entity from the user point of view. Another way of
putting this is that each atomic object has a globally unique *distinguished
name* used to identify it within operations. The definition of an atomic
object is determined by what is the most convenient view for the user.

2.5.1   Attributes

Real world objects may have many properties of interest to users. For example, this document has a title, author, sections, subsections and diagrams.
Properties are represented by *attributes* within atomic objects. An attribute
has a *type* which allows it to be referenced apart from other attributes of the
object and a *value* which represents the specific property of the object. Our
document example includes an attribute with type *author* and value *Steve
Benford*. This paper will use the general notation *type = value* to represent attributes (e.g. *author = Steve Benford*). An atomic object may have
many attributes with the same type (a document might have many authors).
Figure 2.2 shows the basic structure of an atomic object. Figure 2.2 distin-

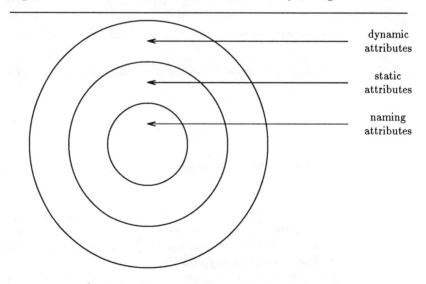

dynamic
attributes

static
attributes

naming
attributes

Figure 2.2: The basic structure of an atomic object

guishes three classes of attribute: *naming, static* and *dynamic.* Naming attributes are used in creating the distinguished name of an atomic object thus the name of an object reflects its properties. Naming attributes cannot be changed once the object has been created which implies that the name of an object cannot be changed (to change a name is to create a new object).

Static attributes (which include naming attributes) cannot be changed once an object has been created except by deletion of the object however, not all static attributes are useful for naming. Examples of static attribute types are *storage date* and *creator* which might be used to hold data concerning the introduction of an object to the storage system.

Dynamic attributes may be altered once an object has been created. New dynamic attributes may be created and old dynamic attributes may be deleted or have their values changed. An example of a dynamic attribute might be *description* which represents a piece of text describing the purpose of an object.

Identifying attribute types

The data model requires a method of identifying attribute types in a globally unique manner. This must insure that each attribute type in the AMIGO information base is globally unique and can therefore be referenced safely. The simplest approach might be to adopt the concept of *Object Identifiers* (OIDs). The AMIGO group could obtain an OID prefix for its group communication framework and all attribute types could be allocated OIDs based on this prefix. This method has the advantage of compatibility with existing administrative policy but is not self contained (i.e. the definition of new types is not an automatic part of the data model and relies on some external authority). However, for the time being, the use of Object Identifiers seems the best approach to take.

Each attribute type is associated with a global indication of whether it is static or dynamic. This mechanism provides a basic integrity constraint on the use of attributes and ensures that static attributes cannot be altered.

2.5.2 Naming

The previous paragraphs have specified that each atomic object has a *distinguished name* which is composed of the *naming* attributes belonging to that object. The distinguished name of an object must be specified when it is created and some mechanism must ensure that it is globally unique. This is facilitated by the adoption of a hierarchical namespace for Information Objects as described in section 2.2.2.

In cases where naming does not occur on a strict organizational basis it is still likely to follow some hierarchy. For example a nested sequence of conferences on different topics could inhabit a tree-like name space where

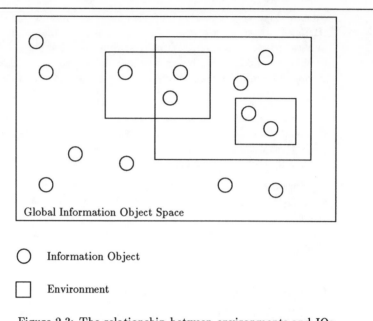

Global Information Object Space

○   Information Object

▢   Environment

Figure 2.3: The relationship between environments and IOs

some topics were subtopics of others.

2.5.3   Environments and Information Objects

Section 2.3 introduced the concept of the *environment* to reflect a binding between entities and Information Objects related by a number of management, access and distribution policies. Each information object may belong to zero or more environments and this is indicated by means of a special attribute of type *environment* whose values are the directory names of environments to which the object belongs. Environments divide the space of Information Objects into logical subspaces which are used to indicate where an object may be found and to limit the locus of data model operations as will be described below. An object belonging to an environment is subject to the various policies associated with the environment.

An environment does not list all of the objects which belong to it although it may indicate the names of *notable* objects which are intended to be globally visible.

Figure 2.3 shows the relationship between environments and Information Objects and in particular, indicates how each object may belong to more than one environment.

This figure also indicates that environments may, in some sense, be nested. The introduction of nested environments produces many new issues to be solved. For example, how are the properties of parent environments inherited by nested environments? These issues have been left for future study.

2.5.4   Operations

This section specifies the operations which apply to atomic objects in the
AMIGO data model. These operations must provide the functionality to
read, modify, create and delete atomic objects. The locus (area of effect) of
many of these operations may be limited by supplying them with the names
of a number of environments which specify the spaces to be searched for the
named objects. For example, the name of an environment may be supplied
to a read operation meaning that only that environment should be searched
for the object. This has practical value because the environment may be
mapped onto a set of *storage agents* to be searched at the internal level.

Read

Read functionality is given by the *Read_object* operation which takes the
distinguished name of an object, a set of target attribute types and a set of
environment names and returns the desired attributes of the named object
having searched the specified environments.

```
set_of_value = Read_object(name, set_of_type, set_of_environment)
```

Modify

The *Modify_Object* operation allows the values of dynamic attributes of a
named object to be replaced or deleted. It also allows new dynamic attributes to be added to an atomic object. It returns error information giving
the final status of the operation.

```
set_of_error = Modify_object(name, set_of_modification,
                             set_of_environment)

modification { set_of {type, add/delete/replace, value } }
```

This operation must enforce static/dynamic attribute constraints for each
specified attribute type to ensure that a static type is not being modified.
The modify operation needs to determine what action to take if some parts
of the operation succeed and other parts fail. For example, if a new attribute
is added successfully but a requested replacement references a missing attribute within the same operation. This paper suggests a simple transaction

control policy of *one out: all out.* This means that if any individual part of a modify operation fails then the whole operation fails and the information base is rolled back to its state at the start of the operation. This approach is chosen on the grounds that it is simple and that users will easily be able to determine the final state of the information base if an error is received.

This operation may be used to add/remove Information Objects to/from environments by modifying the *environment* attribute.

Add/Delete

The *Add_object* and *Delete_object* operations add and delete atomic objects from the information base. Addition of an object verifies that the new name is globally unique and allows some initial attributes to be added to the object (including the names of its initial environments). Deletion of an object removes the name and all associated attributes from the information base.

```
set_of_error = Add_object(name, set_of_attribute)
```

```
set_of_error = Delete_object(name, set_of_environment)
```

2.5.5　Structured Attribute Values

Section 2.5 describes an atomic object in terms of a set of attribute/value pairs. This structure does not offer enough functionality to represent many of the real world objects which we are interested in. If we consider an X.400 IP-Message we can see that it consists of a set of header attributes and a set of body parts where each body part may represent another encapsulated X.400 message. In a real world X.400 IP-Message the attributes representing the encapsulated body part are grouped together within the main message. This grouping should be reflected by the atomic object representing the message within the AMIGO data model.

The desired functionality can be achieved by structuring attribute values. Complex attribute values can be constructed as *Sets* and *Sequences* of simple attribute values which are defined as before. The terms *Set* and *Sequence* are analogous to the *ASN.1* constructs of the same name and refer to unordered and ordered groups of attribute values respectively. These constructors may be nested arbitrarily. The terms *static* and *dynamic* apply to simple attribute values only but otherwise have the same effect as before. Figure 2.4 is an ASN.1 definition of an atomic object.

```
AtomicObject ::= SET OF Attribute

Attribute ::= SEQUENCE {
            AttributeType,
            AttributeValue
      }

AttributeType ::= SEQUENCE {
            type OID,
            constraint INTEGER {                    10
                  static(0),
                  dynamic(1)
            }
      }

AttributeValue ::= CHOICE {
            [0] SET OF AttributeValue,
            [1] SEQUENCE OF AttributeValue,
            [2] SimpleValue
      }                                             20

SimpleValue ::= ANY
```

Figure 2.4: Definition of an atomic object

2.5.6 Examples

The following are two examples of real world objects represented as AMIGO atomic objects.

An X.400 Message can be represented as an atomic object with attributes representing the P2 header information and the body parts. An encapsulated message can be represented by an attribute of type *encapsulated message* with a complex value. Simple body parts such as pieces of text can be represented by attributes of type *bodypart* having a simple value.

```
country: England  -- naming attribute
organization: Nottingham -- naming attribute
type: IP-Message
ID: IP-Message-ID -- naming attribute
environment: name of environment A
recipients: sdb@cs.nott.ac.uk
recipient: jpo@cs.nott.ac.uk
....
bodypart: Hello how are you, I have forwarded ...
encapsulated message
type: IP-Message
ID: IP-Message-ID2
recipient: joe@xx.yy.zz
```

```
. . . . .
bodypart: I've got a great new idea ...
```

A document can be represented by a similar structure, with chapters, sub-chapters and diagrams represented by structured attribute values.

```
country: England -- naming attribute
organization: Nottingham -- naming attribute
type: Document
title: What I did on my holidays-- naming attribute
author: Steve Benford -- naming attribute
environment: name of environment A
environment: name of environment B
section:
heading: chapter1
section:
heading: 1.1
bodypart: This paper ...
bodypart: This section ...
section: ...
. . .
```

These examples emphasize the use of structured attribute values to represent the structure of information in an object.

2.6   Sets of objects

The previous section described the *atomic object* as a method of representing a basic real world object. This section will describe a method of describing sets of objects in a flexible and dynamic way. A set description method is required for the following reasons:

> Many real world objects are sets of other objects. For example, a *conversation* can be defined as a set of messages recursively in reply to each other. The elements of these sets are individual objects in their own right but users will view the entire set as a single object for specific purposes (e.g. read a conference). The *atomic object* mechanism is unable to describe sets of objects where individual members lead an existence of their own.

> Grouping objects allows users to structure information in meaningful ways and manipulate it more effectively. Such group-ings might be highly dynamic, only existing for the duration of

an operation. For example, a user might search the information base for all documents sent by another user. This search can be viewed as the definition of a short term set. Searching and structuring information in flexible ways is important in an environment where users have to deal with a large volume of information.

The set description mechanism should react gracefully to changes in the information base (e.g. A new message in the information base should automatically be included in all relevant sets) and should also reflect the idea of objects existing within spaces defined by environments. We will first consider the problem of representing well known sets which have relatively long lifetimes (these are called *compound objects*) and will proceed to discuss dynamic sets and searching techniques.

2.7 Compound objects

A *compound object* represents a real world object which is a collection of other objects. Compound objects have two parts: a *description* and an *object set*. The *description* assigns a name to the compound object and specifies some properties of the set as a whole. The *object set* consists of all objects which currently belong to the set. For example, the contents of a conference are a compound object: Its description might include a name and attributes describing its purpose, creator and subject. Its object set will be the set of all messages currently belonging to the conference.

The description is represented by an atomic object whose attributes represent general properties of the set. The name of the compound object is the name of the atomic object representing it thus compound objects inhabit the hierarchical name space described in section 2.2.2.

The object set is obtained dynamically from a *derivation rule* which is a special attribute of the description of the object. The derivation rule specifies properties of and relationships between the objects in the object set. For example, the derivation rule for a conference might specify all messages with the subject field having the value *UNIX users.*

The object set is derived from a derivation rule by the *ennumerate* operation which determines all objects satisfying the rule. This operation returns the names of the objects belonging to the set given the name of the compound object containing a derivation rule.

The ennumerate operation may be passed the names of a number of environments. These limit the operation such that only the names of objects which satisfy the rule and are contained within one of the environments are returned. It follows that a compound object may have different contents when viewed from different environments.

Figure 2.5 shows the global information space containing a number of Information Objects and environments.

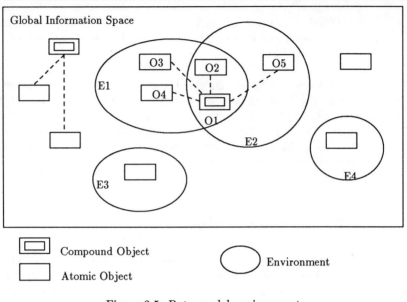

Figure 2.5: Data model environment

If compound object *O1* is ennumerated in environment *E1* its contents will
be *O2,O3* and *O4* whereas if it is ennumerated in *E2* its contents will be it
O2 and *O5*.

```
set_of_names = ennumerate(name, recursion_flag,
                    set_of_environment)
```

The ennumerate operation might return the names of other compound objects and it should be possible to ennumerate these recursively if desired.
For example, the object set of a conference might contain objects describing
sub-conferences and it should be possible to ennumerate these when ennu-
merating the containing object set. The recursive action of this operation
may be switched on or off by a flag which is one of its arguments. An implementation of recursive ennumeration must deal with cases where objects
contain each other in a looped manner. This may be tricky in a distributed
environment.

The *read, modify, add* and *delete* operations which manipulate atomic objects apply to descriptions of compound objects and can be used to create,
delete and maintain compound objects. It is important to understand the
difference between the *read* and *ennumerate* operations. The former returns
attributes from a named atomic object. This atomic object may be the
description of a compound object. The latter returns the names (not at-

tributes) of the object set of a compound object. It does this by expanding
the special derivation rule attribute belonging to the objects' description.

2.7.1   Derivation rules

The *derivation rule* is fundamental to the realization of compound objects.
Derivation rules should allow a great deal of flexibility in defining sets of objects and the following paragraphs describe some of the functionality which
should be supported.

The derivation rule for a set should support a simple listing
of the names of objects contained in the set. These names might
also be the names of subtrees in the name space indicating that
all objects in the subtrees are included in the object set. The
listing of names corresponds to the creation of logical pointers
to objects in a set. It should be noted that when the *delete*
operation deletes an object it does not change any derivation
rules which name the object although the object will not be
present in the object set when it is next ennumerated. Special
operations should be included to explicitly add and delete the
names of Information Objects to and from lists in derivation
rules.

The derivation rule should support the matching of attributes
within objects. Attribute matching can be specified by target
attribute/value pairs linked by the *and, or* and *not* operators.
For example, the rule: *subject = UNIX users AND country =
UK* might describe all objects belonging to a UK wide UNIX
users group. Derivation rules should allow the comparison of
attribute values using the $>, <, =$ operators.

Derivation rules should be able to express the relationships
between objects which belong to the object set. Relationships
can be expressed in terms of attribute/value pairs and can be
linked by the logical operators as described in the previous paragraph. This functionality should apply recursively if necessary
allowing the expression of recursive relationships between messages. For example, a *conversation* could be defined by a rule
expressing recursive *in reply to* relationships between messages
starting at some initial message.

2.7.2   Searching

The second motivation for introducing a set description mechanism into the
data model is to aid users in searching the information base in order to locate the information they require. Searching involves the specification of a

search rule describing a set of objects. A search rule fulfills the same role as
a derivation rule but is dynamic in nature. Thus we can view a search rule
as a derivation rule which is not associated with a named compound object.
Search rules need the same functionality as specified for derivation rules in
section 2.7.1 above. Searching involves the introduction of the *search* operation which ennumerates the rule and returns the names of the objects the
user was searching for. The search operation closely resembles the *ennumerate* operation being constrained by environments and operating recursively
if desired.

```
set_of_names = search(derivation_rule, recursion_flag,
                      set_of_environment)
```

Examples of search rules are:

```
type = message AND recipient = steve@cs.nott.ac.uk
```

which would find the names of all messages received by Steve or

```
type = document AND creation date > 5/12/87
```

which would locate all documents written after 5/12/87.

2.7.3   Deletion from compound objects

In some cases it may be necessary to remove a specific object from a set of
objects without deleting it globally and removing it from all relevant sets.
This might be achieved in the following ways:

- If the set was described in terms of a list of names then the name
  of this object could be removed from the list by means of special list
  manipulation operations.

- The derivation rule for the set could be changed so that this object no
  longer satisfied it. However, it could be difficult to determine how to
  modify the rule when removing one specific object from a large set of
  similar objects.

- The attributes of the object could be ammended so that it no longer
  satisfied the derivation rule. One way of doing this is to specify the
  set derivation rule with an attribute matching clause such as *invalid
  != XXX* and then to add an attribute of the form *invalid = XXX* to
  the object to be deleted from the set. The value of XXX could be the
  name of the set from which the object was to be deleted.

2.7.4   Examples

The following are examples of compound objects. In each example the
descriptive part of the object including its derivation rule are presented.
Each derivation rule could be used as a search rule to find the set of objects
on a one off basis.

A conversation

A conversation can be defined as a set of messages recursively in reply to
some initial message. The following is an example of a compound object
representing a conversation:

```
country: England -- naming attribute
organization: Nottingham University -- naming attribute
type: conversation
subject: next meeting -- naming attribute
environment: name of environment A
creator: ...
...
derivation rule: recursive(in reply to = IPM-ID X)
```

A conference

A conference can be thought of as a set of messages, grouped by subject,
together with a set of entities which access the conference (this interpretation is open to debate). Users actively contribute and subscribe to conferences of their choice. The distribution and storage agents associated with
the conference are specified by its environment description and its contents
are specified by a compound object. Several conferences may belong to the
same environment but each will have a different compound object representing their contents. The compound object representing a conference might
resemble the following:

```
country: England -- naming attribute
environment: AMIGO conferences
conference name: Data model discussion -- naming attribute
type: conference
subject: UNIX users
...
derivation rule: type = message AND
  subject = UK UNIX users AND recipient = xxx@xx.xx
```

The P2 subject field must be used to identify the conference under 1984 X.400; 1988 X.400 will allow special protocol specific fields to be defined for this sort of purpose.

2.8   Access controls

This section will describe the need for access controls within the data model and present a basic outline of a suitable access control mechanism.

Access controls specify which users can perform which operations on which data and are required within the data model for two main reasons:

- There is a requirement for *privacy* within an environment where information is shared between many users belonging to different organizations. Privacy is the right to prevent people from retrieving information or to hide it so that they are unaware that it exists at all. Privacy is required when dealing with personal or sensitive information.

- There is a requirement for *integrity* of information which generally means that information should be correct and meaningful. Integrity is difficult to guarantee but an access control mechanism can assist integrity by ensuring that only authorized people can maintain information. Authorized people should understand the consequences of their actions and should have no interest in deliberately destroying the integrity of information.

The following sections briefly consider the problems of describing sets of users and operations within access controls.

2.8.1   Information and levels of access control

The access control mechanism must control access at the environment, object and attribute levels within the data model. Control over individual objects is clearly needed, for example, to specify who can delete a message or document. Control over specific attributes within an object is also required if the data model is to support the concept of *private* information. Private information might be the addition of an attribute to an object which can only be seen by a select group of users. For example, company XYZ might add a comment to a message indicating that it suggests an idea to be included in a new product. This comment should only be visible to certain employees of XYZ.

Access controls within the data model can be achieved by associating *Access Control Lists* (ACLs) with objects and attributes where an ACL assigns sets of access rights to sets of users. An object ACL assigns rights for manipulating the object as a whole (i.e. for reading it, deleting it, adding new attributes and changing its access controls). An attribute ACL assigns rights for manipulating a specific attribute within an object (i.e. for reading, deleting it, changing its value and changing its ACL).

Access controls assigning the right to create a new object are given by an access control list associated with the new object's superior node in the naming tree.

When a user requests an operation on an object the object level ACL is checked to see whether the user is allowed to know of this object. Attribute ACLs for affected attributes are also checked to determine whether the user has the rights to perform the operation.

2.8.2   Access rights

Permission to perform operations is indicated by the possession of *Access Rights*. Different combinations of access rights are required to perform different operations and a list of access rights is given in the following paragraphs.

- The *detect* access right allows the existence of an object or attribute to be known. If this right is missing any references to the name of the object or attribute will result in an error.

- The *read* access right allows information to be read via a *read_object* operation. If present in an attribute ACL it allows the values of the attribute to be read, if present in an object ACL it acts as a default for all attributes belonging to the object.

- The *create* access right applies to object level ACLs and allows the creation of new child objects in the name tree.

- The *delete* access right allows the deletion of information. If present in an object ACL it refers to the entire object and its attributes if present in an attribute ACL it refers to a specific attribute.

- The *add* access right allows the addition of new attributes to an object.

- The *replace* access right allows the values of attributes within an object to be replaced.

- The *replace acl* access right allows the replacement of the ACL where it is specified.

Table 2.1 shows which access rights affect which data model operations.

2.8.3   Users

The access control mechanism requires a method of describing users so that they can be assigned *access rights* within ACLs. Users can be represented by their *directory names* which will be maintained in the directory service

operation	relevant access rights
read_object	detect read
modify_object	detect add delete replace
add_object	detect create (both at superior)
delete_object	detect delete
ennumerate	detect
search	detect

Table 2.1: Operations and access rights

and the simplest method of representing groups of users is to list groups of directory names. The names of subtrees in the *Directory Information Tree* (DIT) could be used to represent groups of users thus removing the need for manipulating long lists of individual directory names. This would also allow access rights to be assigned on an organizational basis as reflected by the DIT.

2.8.4   Amending the data model

The preceding sections outlined the structure of Access Control Lists which assign access rights to groups of users. The following ASN.1 definitions define the structure of ACLs and note changes to the definitions of objects and attributes due to the inclusion of this access control mechanism.

Data model operations need to be changed to allow the manipulation of access control lists. The *read_object* operation must return ACLs on request and the *add-object* and *modify_object* operations must allow ACLs to be created and replaced respectively.

2.8.5   Access controls affecting sets

This section discusses the affect that the access control mechanism has on *compound objects* and *searching* as defined in section 2.7.

Access controls apply to the *description* of a compound object as they would with any other atomic object thus the access control model governs the creation and deletion of compound objects. The *ennumerate* and *search* operations return the names of objects in a set. These operation should check whether the user has the *detect* right at each of these objects and if not the name should not be included in the final set. In this way the privacy of objects is maintained by the set description mechanism.

```
AtomicObject ::= SEQUENCE {
        ACL,              -- object level ACL
        SET OF Attribute
}

Attribute ::= SEQUENCE {
        ACL               -- attribute level ACL
        AttributeType,
        AttributeValue
}

AttributeType ::= SEQUENCE {
        type OID,
        constraint INTEGER {
                static(0),
                dynamic(1)
        }
}

AttributeValue ::= CHOICE {
        [0] SET OF AttributeValue,
        [1] SEQUENCE OF AttributeValue,
        [2] SimpleValue
}

SimpleValue ::= ANY

ACL ::= SET OF AccessControlElement

AccessControlElement ::= SEQUENCE {
        accessrights BITSTRING,  -- rights are bit patterns
        users SET OF DirectoryName
}

DirectoryName -- defined in X.500 standards
```

Figure 2.6: Data model Access Control Lists

2.9   Conclusion

This chapter has outlined an abstract data model for describing information objects. In particular, it has considered the structure of objects, how they are named and how the relationships between objects may be expressed and manipulated. Another key issue covered by this chapter was the environment which forms the important link between groups and their information.

The purpose of this work was to provide a tool for modelling group communication in terms of the shared access of Information Objects by Communication Entities. A wide variety of group communication structures can be modelled using this tool.

The following chapter will describe a specific realisation of this model within a distributed computer system. It will derive an architecture supporting the cooperation of OSI applications to achieve group communication, based on the above model.

Chapter 3

A Group Communication Service Architecture

Authors:	Karl-Heinz Weiss
	Manfred Bogen
Contributors:	Steve Benford
	Hugh Smith

3.1   Introduction

Chapter 1 established the need to support group communication processes within OSI environments. However, this type of communication effectively requires the concurrent utilisation of several quite separate OSI application systems - including the Message Handling System (MHS), Directory System (DS) and an Archive System (AS). An impossibly complex task for a potential user without some form of of aiding. What is required is a kind of meta-level user interface service that can, where necessary, communicate with these other services on the users behalf. Therefore, a new application system is proposed herein that can integrate and unify the above communication services - the Group Communication System (GCS). (See Figure 3.1.)

In chapter two, an abstract Data Model was presented that describes group communication entities and their activities in terms of operations on Information Objects. (Examples of Information Objects are messages, documents, conferences and conversations.) This provides a framework to allow the specification of the GCS service to be developed without becoming embroiled in implementation issues,

In this chapter we shall describe the conceptual architecture of the Group Communication System (GCS). We start with an overview of the architectural issues. The requirements of three classes of GCS users are examined: end-users of the GCS, the administrator/providers, and designers. Next, the functionality and basic services provided by the GCS are identified. This is

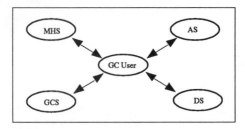

Figure 3.1: User Environment

followed by a discussion of the relations between the GCS and supporting services. Finally, a brief overview is given of the usage of the GCS in a sample group communication application.

3.2   Communication Architecture requirements

In this section we will discuss how group communication activity may be mapped onto a communication architecture model which includes the Group Communication System as *one* component. The communication architecture model will be reflected in the GCS's knowledge about Information Objects (IOs) and communication entities: where the desired IOs are, how they are interpreted, and which communication entities have to co-operate to provide a service. For example, assuming that there is a group IO distribution service, the communication architecture model will define where the appropriate IOs (e.g. distribution list descriptions, user messages) are stored and which external systems (e.g. MHS) and internal functional entities have to be enabled to support the service.

This chapter is primarily concerned with the conceptual and logical structure of the architecture rather than detailed internal descriptions. It outlines a model for describing server configurations and their interaction in order to provide the required functionality to support particular kinds of group communication processes. The discussion of the architectural model is restricted to the application context in the application layer of the ISO/OSI-Model.

The architectural model takes into account three types of constraint: the need to support 'distributed' group communication, the need to support high level descriptions of communications activity, and the need to support the dynamic nature of group communication.

Distributed Communication   The communication architecture model must allow the description of group communication systems that support multiple co-operating entities in distributed environments. The scale of distribution should include office environments as well as international working groups. In the case of groups distributed across organisations and countries, the service will have to be achieved by a 'service molecule', consisting of atomic functional service entities, having different molecule structures like chains, trees or nets.

To meet this requirement a global naming scheme is necessary to identify communication entities and Information Objects uniquely. A further requirement is that the model should support distributed storage - to avoid communication overhead in a very large distributed environment,

Group Activity Specifications It must be possible to support the specification and regulation of existing patterns of group communication activity as well as be able to add new activity descriptions. This implies being able to specify such group activities as 'conferencing' as well as more procedure-oriented activities such as 'editing a report'.

The AMIGO Activity Model [DPB88] provides one approach to the specification of group communication activities. The architecture model described in this chapter provides only the basis for integration with the Activity Model by including the components: Roles, Rules and Functions. This means that entities which can act as Activity Interpreting Entities exist within the architectural model. In addition to the handling of group IOs by specific operations (submit, distribute, deliver, store, retrieve, manipulate), these entities must support the handling of the Role, Function and Rule Component within Activity definitions.

Dynamics of group behaviour A group has several dynamic characteristics; new members can join or leave the group; attributes of group members like addresses, tasks, rights or obligations can change over the course of time. In fact the characteristics of a user can be considered as loose bindings which are defined for a certain period of time, but which can always be modified.

This implies that configuration of the GCS must also be dynamic. This means that, for example, an Archive System agent should not be permanently bound to one group. (The binding between supporting communication entities, such as Archive agents and GCS agents, IOs and users is related to the Environment concept defined in chapter 2.)

We will now turn to the issue of specifying service requirements. A group communication service needs to handle many kinds of requests from different types of user. It is helpful to divide all the required services into sets related to the user's, the provider's and the designer's view. We use the term 'user' in this context to mean an individual who might be involved with different organisations whilst playing different roles, each role having a set of communication related rules [DPB88].

3.2.1   User's view

A User can be an individual, a group member, a group or a subset of a group as shown in Figure 3.2.

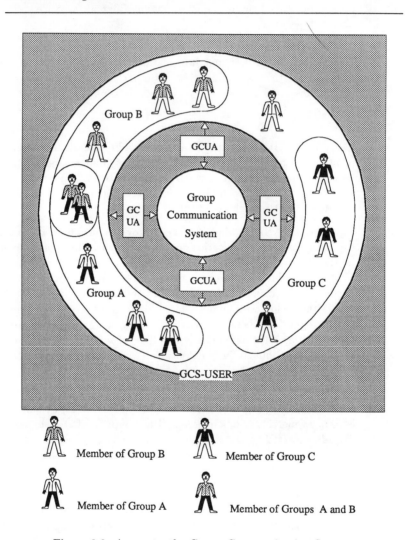

Member of Group B   Member of Group C

Member of Group A   Member of Groups A and B

Figure 3.2: Access to the Group Communication System

A Group Communication Service user requires:

- Interpersonal messaging, to communicate with other users.

- Registration with the GCS for different group communication facilities.

- Communication support services to interact with groups of users. This implies the handling of group IOs at the local interface to the GCS (i.e. collect, create, distribute, filter, moderate, modify, read, search, retrieve, ..).

- Filing and retrieval of documents, to allow location and application independent filing and multi-key retrieval of documents.

- Directory, to know where and how to access remote communication elements, applications or users.

- Encryption, as defined in X.400 (1988).

- Authentication, to avoid unauthorized access.

- Asynchronous (store and forward) as well as synchronous communication mechanisms.

- Information object transfer between different applications or remote servers (third party transfer).

In principle the GCS should adapt to the user's needs whilst taking into account the local environment. The autonomy of a user, on one side, and the needs of groups covering several autonomous environments, on the other side, are reflected in the distinction between the local and global handling of group IOs.

Local handling, in the sense of personal/private or organisational users [Mal85], should be supported by appropriate structured IOs and operations that allow information filtering for local information management. Group information objects could be viewed and structured by a user controlled 'information lens' [MGT+87].

A global shared user-view of group IOs (e.g. multi-organizational or international) must be agreed and defined by the group. Tools are needed to support the global handling of group information objects such as automatic or human moderated conversations.

3.2.2 Administrator's and provider's view

Administration operations are separated from normal user operations. Consequently, a special *port* should be defined for the administrator. (This seems preferable to controlling the whole list of operations via access rights.) The administrator has to prepare, inspect and control the group communication system with regard to administrative agreements and arrangements.

The administrator/provider offers:

- Distribution, configuration and administration of the system components.

- Presentation of the overall service.

- Performance of special processes with high priority and authentication procedures.

- Maintenance of availability.

- Guarantee of service quality.

- Error handling and recovery.

- Registration, analysis and tracing of the system.

- Maintenance of the system as far as efficiency, synchronicity and consistency are concerned.

3.2.3   Designer's view

The system designer requires tools to control the present state of the system in order to get information and feedback for future system development.

In particular the designer needs tools to:

- describe components.

- to analyze the state of a component.

- partially analyze system behaviour.

- extend the system in a flexible manner.

- enable evolution.

Summary   The architectural model that supports the Group Communication Service has to reflect the criteria discussed earlier and the requirements outlined in the three sets of views above. We shall now turn to specification of such a model and discuss the Service and the definition of the components within the GCS.

3.3   The Group Communication System (GCS)

The services offered by the GCS system may be categorised into Basic Services and Advanced Services.

Basic Services are realized within the GCS i.e., a mapping to functional entities is possible. Advanced Services are intended for handling some of the more detailed group activity procedures defined in the AMIGO Activity Model [DPB88]. This chapter mainly considers Basic Services.

The GCS Basic Service is made up of a **Coordination Service**, a **Distribution Service**, and a **Archive Service** - see Figure 3.3. Access to the

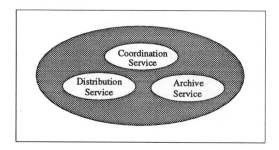

Figure 3.3: Basic Group Communication Service

Service is by means of the concept of a *port* (see Appendix A.1). A port is a point at which an abstract object interacts with another abstract object. Ports are of different types, the type determines the kinds of interactions they enable. An asymmetric port has two instances, *consumer* and *supplier* reflecting a client-server relationship. In the architectural model a Group Communication User Agent (GCUA) accesses the Basic Service ports via an interaction with a Group Communication System Agent (GCSA).

Figure 3.4 shows the proposed GCS port specification. This contains six ports, two related to distribution access (i.e., MHS), two related to Archive access, and one each for coordination and administration. The following paragraphs describe specific aspects of these ports in greater detail.

The Basic Service allows a client to access Information Objects via either a distribution type service - using the *distributionDelivery port* and *distributionSubmission*, or an archive type service - using the *retrieve port* and *store port*. Thus, the Receive operation of the *distributionDelivery port* can be considered to be functionally equivalent to the ReadObject operation of the *retrieve port*.

If a user wishes to selectively accept messages from the Distribution Service he/she can utilise a filter expression. Thus, the GCUA, through the *distributionDelivery port*, can invoke the SetFilter operation to store an attribute expression within the GCSA. This expression is used to select the Information Objects that can be delivered to that particular GCUA.

Similarly, the GCUA, through the *distributionSubmission port*, can perform the SetFilter operation to selectively disseminate messages being inated to all GCUA members of this distribution activity. However, this operation will be subject to access rights, since it is related to the role of moderator or editor of the group communication activity.

The GCUA will also be able to perform the ReadFilter operation in both

administration **PORT**
   CONSUMER INVOKES {
       RegisterGCUA, De−registerGCUA,
       Modify, SetFilter, ReadFilter, List}
   ::=id−pt−administration

co−ordination **PORT**
   CONSUMER INVOKES {
       Answer, AccessRightsControl,
       CreateActivity, KillActivity}   10
   SUPPLIER INVOKES {
       Request, Result}
   ::=id−pt−co−ordination

distributionSubmission **PORT**
   CONSUMER INVOKES {
       Send, SetFilter, ReadFilter}
   SUPPLIER INVOKES {
       Filter}
   ::=id−pt−distributionSubmission   20

distributionDelivery **PORT**
   CONSUMER INVOKES {
       SetFilter, ReadFilter}
   SUPPLIER INVOKES {
       Receive, Filter}
   ::=id−pt−distributionDelivery

retrieve **PORT**
   CONSUMER INVOKES {   30
       ReadObject, AddAttribute,
       DeleteAttribute, SetFilter,
       ReadFilter}
   ::=id−pt−retrieve

store **PORT**
   CONSUMER INVOKES {
       CreateObject, DeleteObject,
       AddAttribute, DeleteAttribute}
   ::=id−pt−store   40

Figure 3.4: Group Communication Port Specification

the *distributionSubmission* and *distribution-Delivery* ports, to allow modification of the existing filter.

To facilitate the restructuring of an archive, the *store port* operations Add and Delete Attribute require a more general definition allowing them to perform attribute modification on a set of Information Objects. As in the enumerate operation, these objects can be selected through a matching expression.

The AccessRightsControl operation, within the *co-ordination port*, is intended to set up the rights and Roles of the group members.

3.3.1 The GCS environment

A user will communicate with a Group via a GCS port. The GCS assumes that the user will utilise the following set of services:

- the DS (specified by ISO/ CCITT X.500),

- the MHS (specified by ISO/ CCITT X.400 '84 and '88),

- the AMIGO Archive System (specified in chapter 4)

- the GCS (specified by this chapter).

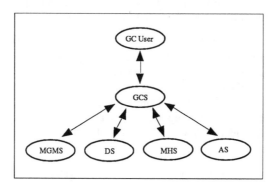

Figure 3.5: Group Communication Environment

To accomplish its function, the GCS *itself* will in turn interwork with these same services and (probably) an additional Management Service as shown in Figure 3.5. To be more precise, the GCS uses the:

1. Interpersonal Messaging Service as an IPM Service User via an origination, reception, and administration port.

2. Message Transfer System (MTS) directly as an MT Service user via a submit, deliver, and administration port.

3. Directory as a Directory User (with or without a DUA) via a search, read, and modify port.

4. An Archive System via a filing, retrieval, and administration port.

3.4   Basic services

Having outlined the architecture and abstract service definition of the GCS Basic Service we will now turn to the discussion of how the functions of the three component services of the GCS (Distribution, Archive and Coordination Services) might be realised. These will be treated in more detail in subsequent chapters (see chapters four to six).

3.4.1   Distribution Service

The Distribution service provides the means of distributing group Information Objects. The service might be implemented by two means. Firstly, by requests to use the MHS Distribution List Expansion Service - this may take the form of normal IP-Messages being addressed to a GCSA agent which has knowledge of, and access to, the Distribution List relevant information. Secondly, Information Objects might be distributed via the Archive system ('pass-through', 'retrieve-to'). The characteristics and advantages of this distribution method as opposed to distribution lists are described in chapter 5.

However, some distributed applications might wish to utilise other Information Objects - e.g. forms, files, ODA documents, SGML documents, sets of documents and communication contexts such as conversations. For the distribution of these IOs, special functionality would be necessary, capable of transferring them in a transparent manner to the application. This might be achieved by co-operation with the Directory Service and the Archive Service.

In order to use the Distribution List service, a user access protocol is required. Furthermore, a system protocol specifying the co-operation between distribution agents must also be defined. In particular, the Distribution service has to be described in terms of abstract operations.

3.4.2   Storage service

At present, there is no external storage system completely fulfilling AMIGO's needs for group communication support. Within CCITT's study group VII, and the corresponding ISO standardization body, a (single-user) message store has been developed as an extension of the former P3 submission and delivery protocol. The design of a multi-user store is being addressed in the activities of ISO/IEC JTC 1/ SC 18/ WG 4 which deal with the specification of a Document Filing and Retrieval Application (DFR). ECMA is also working in this area, having started its 'Document Filing and Retrieval Service Description and Access Protocol Specification'. (See appendix A for more detail on DFR.) Although DFR has many capabilities, it fails to satisfy the needs of an AMIGO Archive Service. The areas in which there is a lack of functionality will be outlined below.

Group Communication requires the storage and retrieval of different group specific Information Objects (not only documents) according to the definitions of the Data Model. Furthermore, the Data Model allows Compound Objects which are used for grouping other IOs. Their behaviour is similar to the 'Groups' defined in the DFR. However, the DFR group mechanism is unable to handle most of the characteristics of Compound Objects. For example, sets, sequences and weak ordering are not supported.

DFR is organized as a single Store. At the present time, there is no support for cooperation *between* stores to meet group communication requirements. In general, distributed Group Communication environments will include several co-operating archiving components.

DFR allows one to define references to documents (to be used instead of the documents themselves), and handles these references by means of internal identifiers, called UPIs. UPIs are not appropriate mechanisms for handling the different kind of relationships needed for group communication purposes. Whereas the Data Model chapter defined a Global Information Naming Scheme for naming information objects within the Group Communication System, the DFR naming scheme defines UPIs for internal use only. Consequently, it is not possible to map the global AMIGO Data Model scheme into the scheme used within the DFR.

Therefore, given the requirements for distributed Group Communication, it is necessary to define an AMIGO Archive Service that fulfills group communication specific needs. As with other services, access to the AMIGO Archive Service is provided via Archive Ports. The Archive Service is accessed via Archive User Agents (AUAs). Due to the range of distribution of communication groups, the Archive System consists of several co-operating Archive System Agents (ASAs).

In summary, the Archive Service provides a range of functions, from those provided by personal user agents (e.g. the Message Store [MHS87b]) to services related to a user group and which may be provided by several distributed and co-operating communication entities.

Finally, as noted earlier, a local and/or global storage and retrieval service, together with user defined information filters for structuring, could serve as a basis for information management within group communication.

3.4.3   Co-ordination service

This service supports the co-ordination of the co-operating communication entities that provide group services. For example, when a user interacts with the GCS via a group communication port, the GCS co-ordination service will then interact with other services to achieve the users goal. Thus the co-ordination service needs to support a number of functions identified earlier in the requirements section, these include:

1. The mapping of 'simple' user requests into a set of operations relating to internal component of the GCS and requests to external services.

2. The co-ordination of the execution of these requests to external services.

3. The registration and management of information concerning users, groups and objects.

4. The support of configuration, error handling and other management processes

An example of the co-ordination of external services would be the distribution of group Information Objects. This might be achieved by the co-operation of several entities: the Directory service defined in [DS87], a group IO transfer service (e.g. the MTS), and an Archive service. Co-ordination is required to set up such services and also to manage dynamic changes e.g. to respond to changes in the user defined filter expressions that enable personalization of the distribution service.

In summary, the basic Co-oordination Service provides the *glue* which enables the Group Communication System to support a range of enhanced communication services.

Co-ordination support for advanced group communication   This report mainly focuses on the definition of group IOs and the low level operations by which they are manipulated. However, an advanced co-ordination service would focus on higher level activities of the kind examined by the AMIGO Advanced Group [DPB88]. Their Activity model can be used as a framework for specifying real world group activities, regulations, roles and functions. Thus, using knowledge of these regulations and the expected behaviour of group members, it would be possible for an activity interpreting entity to automatically create and submit messages according to pre-defined events. Such co-ordination might considerably reduce the 'information flow overload' problem for users.

Some consideration of the advanced activities to which this model might be put is given in chapter eight.

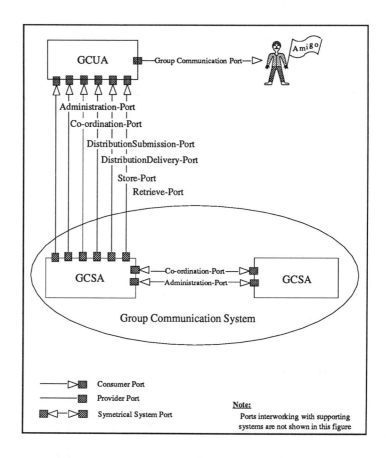

Figure 3.6: The Group Communication System and Service

3.5 The Group Communication System Agent

The central functional entity within the group communication system (GCS) and the representative of the GCS to the outside world is the group communication system agent (GCSA). A GCSA is accessed via a group communication user agent (GCUA). GCSAs and GCUAs have the following capabilities and characteristics:

- A GCUA is an entry point for the GCS which is reachable via a Group Communication Port as shown in Figure 3.6.

- A GCSA is able to co-operate with other GCSAs to provide the required service via a GC system protocol.

- In addition to this co-operation, a GCSA has to co-operate with the following (see Figure 3.7):

    a) the Directory Service (DS),

    b) the Archive System (AS), (AMIGO Multi-User Storage System for piloting)

    c) the Message Handling System (MHS '84 and '88) with message store (MS), (MHS 1984 with AMIGO-DLs for piloting)

    d) the management system (MGMS), to be based on the emerging ISO/ OSI common management information system (ISO 9595 (service)/ 9596 (protocol)),

Cooperation occurs via service specific user agents (e.g. MTS-UA, DUA, AUA, ...) which are integrated in the GCSA.

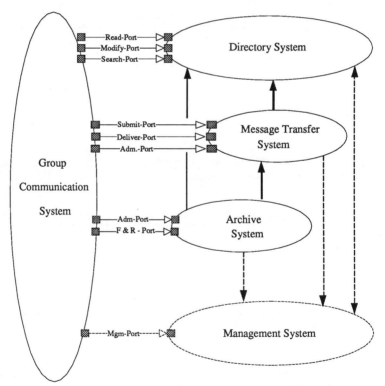

Figure 3.7: GCS interworking with supporting systems

The GCSA enables and controls the access to group-relevant information objects. A user can only achieve the full group communication service via the GCSA. However, parts of it can be requested directly from single functional entities.

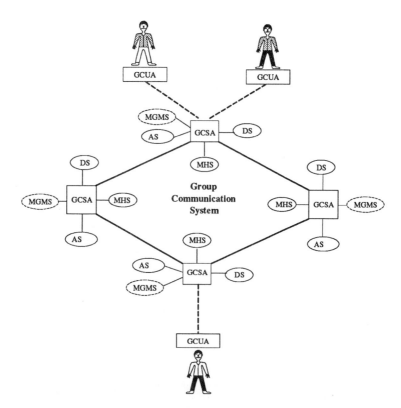

Figure 3.8: Functional Model of the GCS

The GCSA – Group relation

Allocating exactly one GCSA to each group may be sufficient where groups are related to an organisation or located within the same (small) country. However, beyond these limits it is unnecessarily restrictive. Consequently, a loose linkage between groups and GCSAs is chosen.

In the case of groups distributed across countries or within a very large country, the group communication service may be provided by several GC-SAs acting in co-operation. If such a **GCSA team** is supporting a group, each GCSA of the team may have alternate GCSA. Alternate GCSAs are redundant and improve the availability of the group communication service.

Besides these capabilities, a GCSA should be able to act as an interpreting entity for AMIGO activities. However, the necessary functionality requires further clarification.

3.6   Relation between the GCS and supporting systems

3.6.1   Co-operation with an external archive system

This section discusses co-operation between the AMIGO group communication system and the external (AMIGO) archive system or the CCITT's message store.

CCITT's message store (MS) [MHS87b] is modelled as an atomic object, acting as a provider of services to an MS-user, and as a consumer of services provided by the Message Transfer System. It is important to note that the MS acts on behalf of only a single MHS end-user (i.e. it does not provide a common or shared multi-user MS service).

Within the GCS, the MS provides support for a Group Communication System Agent (GCSA), which uses it via a retrieval, submission and administration port. It can also be used for group communication purposes. However, the binding of an MS to a group or a GCSA has to be defined and must always be changeable.

The AMIGO Archive System (chpater four) provides an abstract Document Filing and Retrieval service and enables a user to communicate with a remote Archive-System-Agent in order to access a remote document archive. The AMIGO Archive service provides the capability for large capacity non-volatile document storage for multiple users in a distributed system (i.e. any number of archive users may use the same archive).

The Archive is aware of documents and groups of documents. Furthermore, internal references between documents in the same archive are possible.

Authentication between the GCS system and the archive system is done, as usual, via abstract bind and unbind operations.

3.6.2   Co-operation with the CCITT/ISO Directory System

Global group information objects and communication entities (users, user-groups, application entities and application processes) have to be named if communication islands are to be avoided. This leads to the usage of a Directory Service, globally available to groups distributed across different operating systems, hardware, organisational and national structures.

Instead of developing a separate AMIGO System with Directory functionality, the existing ISO/CCITT Directory System will be adopted.

The usage of the proposed DS standard for group communication is limited to the standard object classes: country, organisation, organisational unit and group of names. However, international communication groups (e.g. MHSnews, english-language conference, unix-user, etc.) may be distributed over several countries and may not be related to these object classes. The use of the Directory, according to group communication specific requirements, is described in chapter 6.

The possible capacity in which the DS can handle extensive usage by the GCS without a significant reduction in performance is an important issue. Consequently, performance issues should be evaluated in the future, probably by early prototype versions of the DS.

In principle it should be possible for each service entity inside the GCS to be a Directory user. Therefore, service independent Directory access is provided to support all functional entities engaged in any group communication activity.

In this context, the DS is utilised to maintain the whole global naming space necessary for group communication which, by its nature, may be globally distributed. Beside this global naming space, local naming subtrees are possible. In some entities, (e.g. in the Archive System), a **local** naming space may be related to the global naming space if the related group has global relevance.

The Directory is accessed by a Directory user via their Directory User Agent (DUA).

There are two possibilities for Directory interaction:

- direct interaction via a DUA (stand-alone) and a DUA interface unit

- interaction via a GCSA with an integrated DUA

The DUA accesses the Directory at its access ports and interacts with it to obtain a service on behalf of a particular Directory user. This access to the Directory utilises the Directory access protocol (DAP).

The Directory itself consists of one or more DSAs, providing access points for a DUA.

The Directory supplies operations via the:

- Read Port (read, compare, abandon), which supports reading information from a particular named entry in the Directory;

- Search Port, (list, search) which allows 'exploration' of the Directory

- Modify Port (add entry, remove entry, modify entry) which enables the modification of entries in the Directory.

The DirectoryBind and DirectoryUnbind operations are used by the DUA at the beginning and end of a particular session with the Directory.

3.6.3   Co-operation with a message transfer system

One approach to the distribution of group information objects is use of the Message Transfer System (MTS). The MTS is a general purpose store-and-forward communication system, supporting all applications of Message Handling. A functional object that provides one link in the MTS' store-and-forward chain is called a message transfer agent (MTA).

The basic Message Transfer service enables MTS-users to submit information objects for delivery and to have information objects delivered to them. MTS-user types, defined in [MHS84b], are: public and private UAs (eg. IPMS-UAs), message stores, distribution-lists, access units or physical recipients.

If an information object cannot be delivered, the originating MTS-user is informed by a Non Delivery Notification. Each information object is uniquely and unambiguously identified. A GCSA (as MTS-user with corresponding interface) specifies the type of the group information object that can be contained in messages which can be delivered to it.

Information object transfer from one GCSA to another GCSA can be achieved using content types other than those defined in [MHS84a] and which are used by IPMS-UAs. To submit group information objects from a GCSA to an IPMS-UA, it is necessary to perform conversions to a format understandable by the IPMS-UA. Those conversions are to be performed by the originator (GCSA).

The content type, the original encoded information type(s) of a message (containing group information objects), and the resulting encoded information type(s) give an indication of any conversions that have been performed.

Access to the MTS Abstract Service is supported by three application-service-elements, each supporting a type of port paired between an MTS-user and the MTS in the abstract model.

The MTS supports ports of three different types [MHS84b]:

- Submission Port (message-submission, probe-submission, cancel-deferred-delivery, submission-control)

- Delivery Port (message-delivery, report-delivery, delivery-control)

- Administration Port (register, change-credentials)

MTS-Bind enables either the MTS-user to establish an association with the MTS, or the MTS to establish an association with the MTS-user. The MTS-Unbind enables the release of an established association by the initiator of the association.

3.6.4 Co-operation with a management system

Management of distributed applications is required to hide network-specific information from human users (transparency) and to minimise administrative overheads and therefore costs. The management user (i.e. the group communication system) should be supported with an optimized resource scheduling mechanism. Additionally, error recognition and recovery are important tasks for guaranteeing the operability of the group communication system.

To meet these requirements, it is necessary to produce, collect, store and administer management information. This information has to be provided for evaluation, (e.g. accounting and billing).

Two entities are responsible for the common management information service: a management system (MGMS) and a management agent (MGMA). The common management information service can be used for advanced network management functions in the following areas:

- Accounting management

- Configuration and name management

- Fault Management

- Performance Management

- Security Management

This service is provided via the management system port. The information for this purpose is provided by the GCS. The resulting directives for the GCS are expected via the MGMA. All functional entities within the GCS are connected with a MGMA.

Currently, a multi-part standard is being developed by the management working group ISO/TC97/SC21/WG4. This standard defines the structure of management information (SMI), and common management information services and protocols (cmis/cmip). This work is at an early stage and therefore the management system integration requires further discussion. Consequently, it is not given detailed consideration within the following work.

3.7 An example: modelling a bulletin board

Chapters 2 and 3 developed a number of models for describing group communication processes. In order to clarify this work, this section presents a brief example of their use to model a simple distributed bulletin board system. It is important to realise that the example supports extremely limited functionality. This section is not intended to model a complete bulletin board system, rather it is included to demonstrate the general modelling principles described by this report.

Section 2.4.2 presented a layered view of group communication. This layered view suggests a top-down approach to the design of group communication systems, yielding the following steps:

1. Specify the functionality of the group communication system in its own terms. This corresponds to the design of the user view and requires the definition of the information and functions visible to various users of the system.

2. Use the Amigo data model to derive an abstract representation of the information objects and operations defining the system. This requires a mapping from the user view to the data model.

3. Specify a suitable system architecture and map the abstract data model to an internal model based on underlying services. It is important to note that there may be many possible architectures utilising different combinations of supporting services. This step involves issues such as the degree of distribution of the system and whether existing services should be used or new ones defined.

The following sections describe the application of each of these steps to the design of a simple bulletin board system.

3.7.1   Bulletin board functionality

The example bulletin board system has simple functionality, consisting of a number of *bulletin boards* on which users post and read *contributions*. The system has two classes of user: *subscribers*, who access the contributions on various boards of interest, and *administrators*, who are responsible for functions such as the creation and removal of bulletin boards.

The following paragraphs describe the information structures and functions available to each of these classes of user.

Information structures

Both subscribers and administrators access two types of information: contributions and bulletin boards.

A contribution represents a single logical item created by a subscriber. Each contribution has a header containing information identifying the contribution and a body containing text. The header consists of a number of fields including, author, title, creation date and subject. In addition, two other fields are of major importance. The *reference* contains a unique system wide numerical identifier for the contribution. The *posted on* field contains the names of the bulletin boards on which the contribution has been posted.

A bulletin board has a header describing general properties such as its name and a textual description of its purpose. The *contents* of the bulletin board consists of the current set of messages which have been posted to the board. Each contribution is posted on one or more boards.

Functions

Each class of user is able to perform a specific set of functions. These are outlined in the table below.

subscriber functions	
submit	post a new contribution to the named boards
read	read a referenced contribution from a named board
browse	show the references of all contributions on a named board
list_unread	show the references of all unread contributions on a named board
list_boards	list the names of available bulletin boards
administrator functions (additional)	
create board	create a new bulletin board
remove board	delete a named bulletin board
remove message	delete the referenced message from the named board

3.7.2 Mapping to the data model

This section describes the mapping between the structures and functions defined above and the information objects, operations and environments defined by the data model. This mapping allows us to view the bulletin board system in an abstract way and therefore aids the design of the underlying architecture.

Information objects

The information structure of the bulletin board system can be mapped to three types of Amigo information object.

A contribution may be represented by an atomic object. In general, there is a direct mapping between the fields of the contribution and attributes. For example, the author of a contribution may be represented by a single *author* attribute. The body of a contribution may also be represented by a single attribute within its corresponding atomic object. The only complication to this mapping is the *posted on* field which contains the names of one or more bulletin boards. This is mapped to one or more distinct attributes, each containing a single name.

The following is an example of an atomic object representing a contribution posted on two bulletin boards.

```
type: contribution
reference: 23
author: Victor Burns
subject: New Big-Moose record
date: 01 01 89
posted on: jazz freaks
posted on: record reviews
body: Has anyone heard the new cut by T.T. B.B. Big-Moose
      Fast-Fingers Hambone Walker? It's fab - particularly
      track 2 on .....
```

A bulletin board may be represented by a compound object. The global properties of the bulletin board may be mapped to attributes in a straight forward manner. The contents of the board may be described by a derivation rule which identifies all contribution objects with the name of the board as a value of the *posted on* attribute. Thus, a bulletin board does not explicitly point to the contributions in its contents. Rather, each contribution points to the boards to which it belongs and the contents of a board are enumerated dynamically.

The following example represents one of the bulletin boards identified within the above contribution.

```
type: bulletin board
name: jazz freaks
purpose: discussion of jazz, blues and soul records
derivation rule: type = contribution AND
 posted on = jazz freaks
```

The third class of information object in the bulletin board data model is a compound object representing index information for unread contributions. For each board a user subscribes to, an *index* compound object contains a derivation rule which can be enumerated to obtain the references of unread contributions. This rule identifies the user, the bulletin board and lists all contributions the user has read on this board (enumerated with the NOT operator to determine unread messages).

The following example index could be used to determine the contributions on the *jazz freaks* board which had not been read by the user *Victor Burns*.

```
type: index
user: Victor Burns
bulletin board: jazz freaks
derivation rule: type = contribution AND
 posted on = jazz freaks AND
 NOT (reference = 2,4,7,9,13,15,16,20)
```

Operations

The functions defined within the user view of the system are mapped onto data model operations as follows:

- The *submit* function is mapped onto the *add object* operation creating a new contribution atomic object.

- The *read* function is mapped onto a *read object* operation, followed by a *modify object* operation updating the relevant index object.

- The *browse* function is mapped onto the *enumerate* operation, applied to the compound object representing the named bulletin board.

- The *list unread* function is mapped onto the *enumerate* operation, applied to the relevant index compound object.

- The *list boards* function is mapped onto the *search* operation. This employs the derivation rule: *type = bulletin board* and operates non-recursively.

- The *create board* function is mapped onto the *add object* operation, adding a new bulletin board compound object.

- The *remove board* function is mapped onto the *delete object* operation.

- The *remove message* function is mapped onto the *modify object* operation, deleting the relevant *posted on* attribute from a contribution object.

Environments

The simplest application of the environment concept to the bulletin board system is to include the entire system within a single environment. Thus, all bulletin boards and contributions inhabit one environment defining the scope of search and enumerate operations. Furthermore, this environment specifies the binding between the abstract data model and the agents and entities specified by the system architecture (see below).

The bulletin board environment is represented by an entry in the Directory service. This entry may contain the names of notable objects in the system.

For example, the directory entry might include the names of all bulletin board objects. In this case, the *list boards* function could be mapped into a Directory query.

The use of a single environment implies that the bulletin board system is unlikely to operate on a large scale. A global system might require the definition of multiple overlapping or nested environments.

This concludes the discussion of the abstract data model for the bulletin board system. The following section specifies a distributed architecture which could support an implementation of the system.

3.7.3   A distributed architecture

The distributed architecture describes the implementation of the bulletin board system by the cooperation of new and existing communication services. It is vital to understand that the abstract data model described above could be mapped onto many possible architectures of which the following is just one example.

The bulletin board architecture specified by this example is based on the general architecture developed within chapter 3. In this model, a new service, the *Group Communication Service (GCS)*, coordinates a number of other communication services to achieve the desired functionality. In particular, the user does not interact directly with these other services. Rather, the GCS acts on behalf of the user. This particular example further refines this model in the following ways:

- Information is acquired by shared access to the distributed *Archive service* and *Directory service* as opposed to a distribution based approach where information is transferred to individual users. Thus, the overall philosophy of the architecture is based on shared access to archived information.

- Communication with the Archive service uses 1988 X.400 as a transport system for operations and results. This is in contrast to a connection oriented approach supporting an interactive access protocol.

- Communication with the Directory service is connection oriented.

The adoption of an archive based approach is supported by the layered system architecture shown in figure 3.9.

The Group Communication Service belongs to the highest layer of the architecture. It is responsible for interpreting user queries, mapping them to data model operations and decomposing these into combinations of queries to lower layer services. For example, a user request to list the names of bulletin boards might be received by the GCS, mapped into a search operation which, in turn, would be decomposed into queries for the Directory

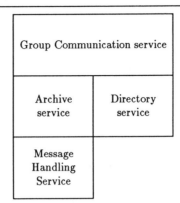

Figure 3.9: Layered system architecture

and Archive services. The GCS would coordinate these queries, collect the results, perform the reverse mappings and return them to the user.

The Archive service (AS) is used to store the contribution, bulletin board and index information objects defined by the data model. In general, the AS receives data model operations from the GCS, implements them and returns the results.

The Directory service is used to store the bulletin board environment description. This description is used by the GCS to determine which Archive agents should be accessed for a given operation. Thus, the Directory provides the information which enables the GCS to coordinate distributed access to Archive agents.

Finally, the Message Handling Service is used to transport operations and results between the GCS and the Archive Service. In effect, both Group Communication System Agents (GCSAs) and Archive System Agents (ASAs) can be viewed as special X.400 P2 entities communicating via the MHS.

In order to clarify the interaction between the GCS, Archive service and Directory service, figure 3.10 shows an example arrangement of cooperating system agents. Each Group Communication System Agent (GCSA) might interact with many Archive System Agents (ASAs). This supports full replication of all information objects between all ASAs. Thus, if a new contribution is posted to a bulletin board, it must be stored within all ASAs implementing the Archive service. To do this, the coordinating GCSA first looks up the names of the relevant ASAs in the Directory environment description. It then creates an X.400 message containing the update, addressed to the ASAs. Consequently, the update is distributed via the MHS. In summary:

- The Archive service stores information objects, fully replicated between ASAs.

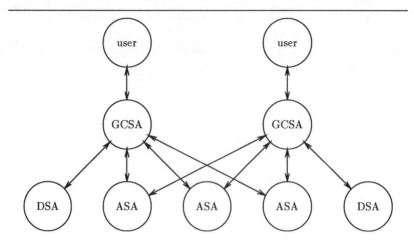

Figure 3.10: Interaction of system agents

- The Directory service provides the knowledge of the ASAs to be queried or updated throughout the course of an operation.

- The MHS provides the transport system for operations between the GCS and AS. In particular it can be used to distribute an operation (e.g. an update) to multiple ASAs.

- The GCS coordinates the interaction of these services and controls the overall execution of operations.

The following section presents some examples of the mapping from user functions to data model operations and their subsequent execution.

Example 1: submitting a contribution

The first example considers a subscriber submitting a new contribution to two bulletin boards. The subscriber contacts the GCS and requests the *submit* function, supplying the new contribution.

The GCSA maps the submit function to the *add object* operation and determines that this requires the update of all system ASAs. Consequently, the GCSA contacts the Directory and reads the X.400 addresses of the relevant ASAs from the bulletin board environment description. The GCSA then creates an X.400 IP-message containing the add object operation. This is distributed to the ASAs via the X.400 MHS. The ASAs perform the update and return the results to the GCSA via the MHS. Finally, the results are collected, mapped to the user model and returned to the user.

Example 2: listing unread messages

The second example considers a subscriber requesting the *list unread* function on a named bulletin board. The GCSA receiving this function maps it to an *enumerate* operation applied to the *index* object for this subscriber on this board.

Due to the full replication of information objects between ASAs, the GCS can direct this operation to any ASA. The name of a suitable ASA can be acquired by a query to the Directory service. The operation is then mapped into an X.400 IP-message which is transferred to the relevant ASA. The ASA performs the operation and returns its results to the GCSA in the form of an X.400 IP-message. The results are then mapped to the user model and returned to the user.

Example 3: listing the names of bulletin boards

The final example considers two approaches to the execution of the *list boards* function.

Firstly, if the names of bulletin boards are included as notable information objects in the environment description, they may be retrieved by a single *read entry* request from the GCSA to the Directory.

Alternatively, the names of bulletin boards can be acquired by a *search* operation directed at any of the system ASAs (due to full replication). As before, the name of a candidate ASA can be retrieved from the Directory service.

3.7.4   Summary

This section has demonstrated an example use of the Amigo models to describe a group communication process, namely a simple bulletin board system. This work has not been intended to describe a complete system. The modelling process occurred in three stages:

- Describe the functionality of the system (user view).

- Map this functionality to the abstract data model.

- Use the data model to derive a suitable system architecture.

The example architecture adopted an archive based approach for the access of information. Examples were given demonstrating the coordination of the Archive and Directory services by the GCS, using the MHS to transfer operations and results.

This concludes chapter 3 describing a general architecture for distributed communication. The following chapters turn their attention to specific aspects of this architecture. In particular, chapter 4 specifies an Amigo archive service.

Chapter 4

Archive Services

Author: Miguel Nuñez

Contributor: Yi-Zhi You

4.1   Introduction

The need for an archive system to support Group Communication has been established in earlier chapters. The primary purpose of the Archive System is to hold and manipulate the Information Objects (IOs) defined in chapter two on the Conceptual Data Model.

Based on the ideas explained in the Conceptual Data Model, this chapter develops the *Internal Data Model* which specifies the methods of representing the conceptual schema in the AMIGO distributed environment. It is concerned with issues such as locating objects, replicating objects, distributed update and consistency. The internal schema considers the properties of services which will implement the distributed storage of information.

This chapter describes how distributed Archive Service Agents (ASAs) may be used to handle IOs and the operations the user is able to use. The Environment concept (which was introduced in Chapter two) is seen here as the basis for distribution of information within different agents. Replication of environments in different ASAs is studied and therefore, the concept of an Environment's Instance is explored.

4.1.1   Overview of the Storage Service

An Environment was defined as a logical structuring of a Global Naming Space. This definition of Environment can be viewed as a logical binding between users and the set of Information Objects that they are allowed to handle.

The Group Communication System uses Environments to carry out the tasks related to group activities. As the Group Communication System uses the Archive System for storage purposes, these abstract Environments must be mapped into the Archiving System Agents which will be the entities that actually hold the Information Objects belonging to the Environment.

To realize this mapping between an abstract Environment concept and real information containers, the concept of an Environment Instance is developed. An Environment Instance is the realization of the abstract Environment concept in actual Archiving Agents.

As the Archive System is distributed, Environments will be mapped by means of a set of Environment Instances distributed in several Archive System Agents. Thus, different Instances of an Environment will contain some part of the Information Objects belonging to it, so the whole information contained within all the Environment Instances is the actual information of the Environment.

This chapter describes the different techniques for handling Environment Instances, in the sense of cooperation between agents to assure consistency of information and update efficiency.

Because the Environment is the basic information container it is also responsible for verifying the uniqueness of names of objects contained within it. Therefore, the Archiving System (AS) is responsible for the consistency of the whole contained information. It is possible that the AS could communicate updates to other external agents (such as UA-Stores). However, the AS will not be responsible for verifying the coherency of information existing in external agents, it only provides tools for the agents to maintain information.

The Internal Information Model

The following paragraphs identify some of the requirements which must be met by the AMIGO Internal Data Model.

- Must represent the basic communication objects defined in the Conceptual Data Model.

- Must represent the information needed for handling distribution of information within the Archive System.

- Must define the elements and operations to support the different policies for handling access to the distributed information and maintenance of replication.

- Must support the multiuser characteristic of the system, and allow mechanisms for Access Control and Privacy of information.

- Must support a namespace of globally unique names which can be used by programs and users accessing information.

- Must insure the Coherency of information in the whole system, and support mechanisms for preserving integrity of updates performed over the system.

Objects in the Internal Data Model

All the Information Objects defined in the Conceptual Data Model have be handled in the Internal Data Model. However, a slightly different view of the IOs must be taken in the Internal Model, because of the influence of distribution concepts.

Within the Archive Service view, an Environment is defined by a name and certain attributes representing information about the IOs it contains. The Environment also contains a list of the IOs belonging to it.

Different instances of the Environment can be duplicates, which contain exactly the same information, or can be partial copies in which only some IOs are held. Therefore, two options are possible for each IO:

- The IO is *Present*, so the whole information of the object is contained in that instance of the Environment.

- The IO is *Absent*. In this case, the instance only contains a pointer to the actual container of the IO. The pointer will contain the Name of the IO and the Reference to the ASA where the proper Instance of the Environment is placed and which contains the IO as Present.

Currently, it is only possible to reference an IO within the same Environment, the possibility to establish references between different environments is left for further study.

Whenever an IO is Present, all the attributes must be present (in this case, it is called "Full Copy" of the IO).

Pointers will only contain the Name and the Reference to the placement of the Full Copy of the IO. Pointers are called "Name Copies" and are composed of:

- Naming attributes

- Access Control Attributes

- Local attributes (for handling of distribution and replication)

Atomic and Compound Objects

Atomic Objects have been defined in the Conceptual Data Model as the basic building blocks which may be used to create compound objects or sets.

Each Atomic Object represents a single real world object such as a document or message. Atomic Objects can be handled by several Environment Instances by means of replication facilities. Thus in each Instance they could be determined to be either present or absent (see previous paragraphs).

Compound Objects are defined as a collection of other objects, Atomic Objects or even other Compound Objects. Compound objects have two parts: a description and an object set. The description assigns a name to the Compound Object and specifies some properties of the set as a whole.

The behaviour of Compound Objects copies are slightly different from Atomic Objects. From the replication viewpoint, the Description Part is similar to an atomic object. However, the Object Set Part has a different behaviour. Atomic Objects belonging to a Compound Object can be either present or absent in certain instance of the Environment. So whenever a copy of a compound object has to be made, it is necessary to specify whether copies of the components will be made or not. Full copies of Name Copies can be used for this purposes.

Because a Compound Object can also contain other Compound Objects, copies of atomic objects belonging to them must be made by specifying the "Depth" of the copy mechanism to be applied to recurrent occurrences.

4.1.2   Techniques for handling Replication of Objects

Techniques for handling replication of information are based on the following principles:

- They depend on the degree of consistency needed.

- The system is only coherent on a long term basis, so updates are performed assuming delays. Temporarily, the system is not fully coherent, so updates are not immediately performed. These are carried out after a certain period of time.

- The Environment is the basis for distribution policies. ASAs provide the physical infrastructure for performing operations on Environments.

The Environment as the basis for distribution of information

The Archive System is organized as a set of Environments placed in different ASAs. (One possibility to get better functionality is to support replication of a certain environment in several ASAs.) The mechanism proposed to get system consistency is based on the Mastership of information.

An Environment is considered by the Archive System to be the collection of the Environment's Instances placed in the different ASAs, so the part of the Environment placed in a certain ASA is called the Instance of that Environment.

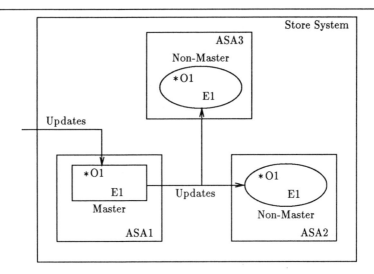

Figure 4.1: Master and Non-master updates

Two kind of instances can be defined: Master Instance and Non-Master Instance of the Environment. The main difference between them is related to the possibility of handling updates. Only Master Instances of the Environment can accept operations for updating the contained IOs.

The Non-Master Instances cannot handle updates by themselves, so updates will be received through other Master Instances. In order to control consistency better, only a few Master instances would normally be defined. This relationship is shown in Figure 4.1.

It is possible to have more than one Master Instance of an Environment; this implies that the actual consistency of the system will suffer some delay because of the time taken by the ASAs to communicate updates. However, although the option of having a single Master is the simplest one, the fact that updates would be carried out by means of a single ASA could constitute a "bottleneck".

There is another difference between Master and Non-Master Instances of an Environment. Non-Master Instances may contain only part of the IOs belonging to the Environment. This is because these Instances are not responsible for the consistency of the information (i.e. they only hold copies of the information). However, Master Instances need to know about all the IOs in the Environment. These Instances do have to recognize duplications of IOs and must be able to handle all the updates to the IOs. Therefore, Master Instances must have Full or Name copies of all of the IOs belonging to an Environment.

The definition of a Master Instance of a Environment implies that all the IOs contained (as Present) within that Instance are Master copies of the IO (i.e. it is possible to handle updates on these IOs). Non-Master Instances of an Environment imply that each object belonging to it is held as a Non-Master copy of the IO. The possibility of handling Non-Master copies of IOs in Master Instances is for further study.

In summary, both Master and Non-Master Instances hold IOs, which can be Present or Absent. In Master Instances all Environment IOs must be either Full copies or Name copies. Non-Master instances may only hold copies of Present IOs and avoid using storage space by holding Name copies. The policy for establishing which IOs are Present or Absent in a certain environment is a local matter defined by the administrators of the environment in the different ASAs.

Updating policies

Because updates can only be performed by users in Master Instances of the Environment, these updates need to be communicated to the different (Master and Non-Master) instances.

There are two choices:

- Sender initiative. Updates will automatically be sent out by the Master which performs the update operation to other ASAs containing Instances. This method is usually very fast, but could imply some overload for the ASA containing the Master Instance.

- Receiver initiative. Updates will be requested by the (Master or Non-Master) Instances which want to be updated. In response to these queries, the asked Instances (Master or Non-Master) will answer by sending the requested updates. Some periodic pooling is involved here and so this strategy will be referred to as the "pooling" policy.

Administrators could choose any combination of these two policies for building the whole behaviour of the environment. An example can be found in Figure 4.2.

- E1 in ASA4 asks ASA3 for updates

- E1 in ASA1 asks ASA2 for updates

- Master E1 in ASA2 send updates directly to ASA3 (updates produced in ASA2)

- Master E1 in ASA2 asks ASA3 for updates (updates produced in ASA3)

The differences between these two policies are mainly related to the degree of consistency and the necessary delay to achieve it.

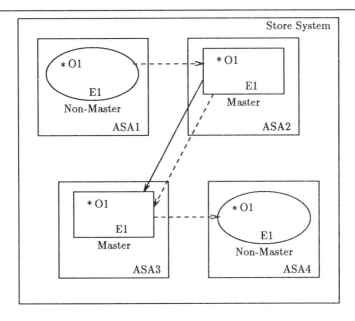

- - -> Ask for Updates (Receiver Initiative)

——> Send Updates (Sender Initiative)

Figure 4.2: Example Administration Policy

Whenever the "sender initiative" policy is used, consistency is guaranteed (between those Instances which automatically receive updates). For example, in the simplest case, whenever a single Master is used and updates are always sent automatically by the Master, global consistency is achieved. The "pooling" policy (receiver initiative) insures real consistency whenever pooling is performed over Master Instances and the time between queries does not tend to infinite.

However, the update delay varies greatly depending upon what kind of policy is used. The global delay for insuring total consistency of the information within an environment is composed by the maximum value of:

- The time spent to transfer updates from the Master Instance to all other instances ("sender initiative" policy).

- The maximum period of time between queries in the case of instances which follow the "pooling" policy.

It is expected that the second quantity will probably be the highest one in every case.

Nevertheless, because consistency is based on the Mastership characteristic of some Instances, where the user actually needs (in cases whenever the requirements of reliability are really high) to contact Master Instances, so the delay must be considered only in terms of Master instances (using the above criterias).

Database Event Records (DER)

Information Objects are Named in the Store Context. Each object has an associated sequence of DERs which is used for distribution handling proposes.

Environment Instances need to propagate between themselves the fact that an event has happened (i.e. events need to be registered within). DERs are representations of Events which happen within the Environment. The main reason to hold this information is to use it in the updating of Environment Instances.

DERs are created by Environments as result of operations performed by the users. DERs are internally handled by the Environment and users do not usually have to be conscious of them. DERs are interchanged in the updating of Environment Instances by means of the proper operations.

The type of operation performed is included as part of a DER. The correlation of DERs to IOs is preserved and a unique identifier for DERs is needed within the Environment.

Because DERs are elements to propagate operations between Environment Instances, they are only useful whenever replication of IOs are allowed. For non-replicated IOs DERs are not needed, so it is unnecessary to use storage space for them. However, because DERs also represent the history 'of operations performed on IO they could even be held in non-replicated Environments if their administration wished.

A DER is composed of:

- DER identification (ASA Name + Local Id)

- Object Reference

- Operation performed (Add, Modify, Delete, etc..)

- Creation Date and Time

- Entry Date and Time

- Attributes affected (pairs of attribute-value)

- Compound Object involved (for compound objects, addition/deletion of new/old documents, etc...)

Although implementation of DERs within ASAs is a local matter, it is proposed that only one database be used for storing DERs in an Environment (instead of having DERs stored in close association with each object). Otherwise one would need to hold deleted objects in order to be able to handle their associated DERs. With a common DERs database one could physically delete the object and after some proper time delete its DERs. One could even just hold the DER of the deletion operation.

The DER unique identifier is assigned by the ASA on a sequential basis. This identifier is composed of the ASA Name plus a local (sequential) number. The ASA is responsible for providing this number. An Environment DER database is able to hold all DERs produced by its own ASA as well as DERs received from other ASAs.

External synchronicity

The AS is responsible for the consistency of its own information contained within ASAs. However, it is possible to synchronise external storage (UA-storage) by means of reading objects and getting DERs. So, the external storage could ask for objects and events produced within them, and hold that information in an updated form.

The AS is not responsible for maintaining the consistency of such information, being only responsible for providing tools that can be used for these tasks. Because reading of objects is always possible by using a normal port, additional functionality (which is mainly designed for internal synchronization) could also be provided that allows the users to get the DERs.

External Agents hold the IOs as local copies, so updates or modifications performed on these Agents are considered local and have no influence on the corresponding IOs in the AS. The only way to carry out modifications on an IO in the AS is by performing the proper operations by means of Archive Agents.

How the Model Works

A user contacts an ASA to perform some operation. The user specifies the name of an Environment within which operations are performed by means of a "Bind_Operation". If the Environment belongs to that ASA the bind operation is completed, otherwise (if the parameters of the operation allow chaining) the ASA contacts the DS and retrieves the correct ASA to contact. Another bind operation is then sent to the correct ASA and the operation is completed. All the operations will be passed through from user to the final ASA. User authentication is performed by the first ASA.

Whenever the Bind is successful, the Environment tries to find the Information Objects referred to within the operations. If those IOs are not present in this Environment Instance, the correct ASA will be accessed (by using the Directory System if necessary) and the operation will be forwarded.

If the IO is found in the present ASA, the Environment Instance could be a Master or not. If a Master Instance is reached, no other problems exist, the operation will be performed and other Instances (in case updating must be initiated by the Master) will be informed by means of the corresponding operation. If the Instance is not a Master, another ASA must be accessed in the case of update operations.

Operations will have parameters for indicating the possibility of chaining or multi-casting of operations. If these possibilities are not allowed, then the operation will return the correct error indication.

4.1.3   Attributes

As defined in chapter two, every Information Object is composed of attributes. Attributes of an object convey various information e.g. properties of the object, links with other objects, etc. Attributes can be used to attach information to objects, to retrieve objects or to organize objects.

Attributes structure

An attribute has a name which identifies the attribute within the containing object. The name of an attribute is used to refer the attribute, in general it conveys the meaning of the attribute for users.

Attributes can represent different aspects of an object, for example, properties such as "InterestDegree", or links such as "InReplyTo", etc. Attributes may be modifiable such as "InterestDegree", or not modifiable such as the "ReplyTo" of a message object. In general, attributes also have properties that can influence the behaviour of attributes. For example, a "not modifiable" attribute cannot be modified after its creation. We call the properties of attributes *qualifiers*.

The value of an attribute gives the exact meaning of the attribute at any given moment. In general, a value type is associated with any individual attribute specifying the possible form or data structure of the values of the attribute.

Naming of Attributes

The only requirement for a name is that it can distinguish the attribute it represents from others in some context.

One obvious solution is to take a globally unique name in some name space, and then the name of attribute will be unique in any context. The proposal is to use ASN.1 Object Identifiers for naming attributes.

Another naming method relies on the the observation that a user works only with a particular set of objects (for example, a conference is concerned only with the objects of that conference.) So an attribute name need only be unique for this set of objects. Typically, a set just contains objects belonging to some application (e.g, all objects in a MailBox). The uniqueness of names is guaranteed by the application in its objects. The only requirement from Environments is that a name must be unique in the same object.

Some Special Qualifiers of Attributes

Some important qualifiers or properties are:

- Name: an attribute is a "name" attribute if it is used to compose (with other name attributes) the name of the object. All name attributes form the distinguished name of the object.

- Modifiable or Not-Modifiable (dynamic or static): A "modifiable" attribute can be modified dynamically after its creation, but the "non-modifiable" attributes cannot. Name attributes are generally "non modifiable".

- Single value or Multiple value: Name attributes can have only one value. However, the "ReplyTo" attribute for example can have many values.

- Be used for retrieval or not: Not every attribute can be used for retrieval. For example, a long text attribute couldn't be used for retrieval.

- Link: A "link" attribute contains references to other objects.

- Global or Local: An attribute is global if it always has the same value(s) for all copies or instances of the object in any AMIGO Environment Instances. For example, the "MId" attribute of IPMessages is global. The non-global attributes are local, i.e. it can have different values in different copies or instances of the same object. For example, the "last modification time" is different for the same object in different Environment Instances. A consequence is that global attributes need to be synchronized in order to guarantee the consistency of different copies or instances of an object, local attributes do not.

- Automatic or Manual: Value of attributes (in particular local attributes) can be either generated automatically by Environments (e.g, the EntryTime) or manually by users.

- Universal: An attribute is universal if it must be present in every object of the Environment. Those last two properties are themselves local to the individual Environment, e.g. an attribute may be universal in one Environment but it is not in any other one.

The above list reflects the most important classes of qualifiers, which are considered sufficient for AMIGO proposes. However, depending on the IOs considered, it may be possible to find other classes.

Value and Value Type

The type specifies the data structure of the value which can be of arbitrary complexity. It can be built by using the constructor "sequence", "set", "choice", ... recursively on other types. The basic types include, Integer, Boolean, Character String, etc.

From the object-oriented point of view, the value of an attribute is an instance object of the class defined by the value type of the attribute.

General Attributes

In this section, we will give some examples of attributes which are independent of particular Environments and may probably be meaningful for AMIGO applications.

- Object Identifier: If there is only one name attribute in the object, the distinguished name of the object shrinks as the value of the attribute "Object Identifier", it is not modifiable, can be used for retrieving, and is in general universal. For the "IPMessage" (a kind of static, atomic object), it is the "MId" defined in the P2 protocol.

- Creator: The user who created the object, the attribute is multiple value, be used for retrieving and not modifiable.

- Create Time: The time this object is created. It is single valued, can be used for retrieving and is not modifiable.

- Derived Rule: This attribute must be present in compound objects, it is, in general, modifiable.

- Access Control: This attribute specifies who can do what operation on the object. It is, in general, universal, modifiable.

Internal Archiving Attributes

In this section, we will define some archive specific attributes, (Note: archive specific means that the value of the attributes are related to the particular Environment. They may be used by every application which uses Environments.) Those attributes are generally local to the individual Environment.

- Owner: the user who owns the object. It is universal, modifiable and can be used for retrieving.

- Entry Time: the date and time at which the object is entered into the particular Environment. It could be, of course, different for each instance of the Environment. It is universal, not modifiable by user (generated automatically by a store) and can be used for retrieving.

- Last Modification Time: the time of the last modification. For a Environment, it seems useful to make it universal. it is also not modifiable by users, and can be used for retrieving. The value of this attribute is generated automatically by store.

- Last Access Time: It has the same properties as the Last Modification Time attribute.

- Size: the size of the object, it may be different in different Environments depending on the coding used in the corresponding systems.

- Local Identifier: This may be used by the individual Environment to make references faster. It is generally generated by Environments.

- Expiry Date: Its value depends to the capacity of the Environment, it is modifiable.

- History: The history of the object, i.e. the list of events. This attribute can be used to perform retrievals such as "the object that I have not read, ...". It is not modifiable directly by users, the value is generated automatically by the Environment according to the operations performed on the object.

Attributes for handling replications

Each object will have an attribute which indicates the type of replication allowed:

- No replication allowed.

- Name replication allowed.

- Full copies allowed.

Users could specify restrictions on the replication of objects by using this attribute.

For Compound Objects, it will also be specified whether copies of the components will be made or not. The "Depth" of this policy to be applied for recurrent occurrence of the Compound Objects also has to be defined.

The general attributes needed for handling copies are:

- Type of Replication. Values are: no copies allowed, Name copies allowed, full copies allowed.

- Is Master. Values are: Yes or No.

- Location of Master (only for Non-Master copies). It will contain the Environment where the Master is located.

- Copies Location and Policy. (Only in Master copies) For each copy, the Master will record the Location of copies and whether updating will be performed by the Master or by the copies using the pooling policy.

- Size of copy. Values are: Full or Name.

- Replication of Components. (for Compound Objects) Values could be: Yes or No.

- Depth of Replication (for Compound Objects) Values could be: "Number indicating level" or "Whole".

These attributes will exist in all copies (depending on the type of copy and object). However, unlike the other attributes, they will not be transferred between the different copies of objects. Thus, they are local to each copy and their values can be different in several copies of an Information Object.

4.2   Storage Protocols

In conclusion we give the following ASN.1 specification of the storage protocols.

—— Useful Definitions

UsefulDefinitions {}
DEFINITIONS ::=
BEGIN

EXPORTS
      protocols, serviceOperations, distributedServiceOperations,
      bindUnbindOperations, module, serviceElement,
      applicationContext, usefulDefinitions, generalArguments,   10
      protocols, serviceOperations, distributedServiceOperations,
      bindUnbindOperations,   *—— to be completed*

ss   **OBJECT IDENTIFIER** ::= {??}

—— Categories of information object ——

module   **OBJECT IDENTIFIER** ::= {??}
serviceElement   **OBJECT IDENTIFIER** ::= {??}
applicationContext   **OBJECT IDENTIFIER** ::= {??}   20

—— modules

usefulDefinitions   **OBJECT IDENTIFIER** ::= {??}
generalArguments   **OBJECT IDENTIFIER** ::= {??}
protocols   **OBJECT IDENTIFIER** ::= {??}
serviceOperations   **OBJECT IDENTIFIER** ::= {??}
distributedServiceOperations   **OBJECT IDENTIFIER** ::= {??}
bindUnbindOperations   **OBJECT IDENTIFIER** ::= {??}
                                  30

END

Figure 4.3: Storage Protocol Useful Definitions

-- General Arguments and Parameters

GeneralArguments {}
DEFINITIONS ::=
BEGIN

EXPORTS
        ServiceControls, ChainingArguments

        -- definitions   10

        ServiceControls ::= **SET** {

                chainingAllowed [0] **BOOLEAN**,
                listSize [1] **INTEGER**,
                useMaster [2] **BOOLEAN**,
                timeLimit [3] **INTEGER**,
                sizeLimit [4] **INTEGER**   }

        ChainingArguments ::= **SET** {   20
                initiator [0] DistinguishedName,
                ssaPath   [1] **SEQUENCE OF** DistinguishedName }

END *-- Module General Arguments*

Figure 4.4: Storage Protocol General Definitions

```
BindUnBindOperations {}
DEFINITIONS ::=
BEGIN

EXPORTS
        SsBind, SsUnbind, SsaBind, SsaUnbind


        -- BIND and UNBIND Operations between SUA and SSA        10

        SsBind  ::= BIND
                BIND ARGUMENT  SsBindArgument

        SsBindArgument ::= SET {
                securityAttributes [0] Credentials OPTIONAL,
                environment [1] ObjectNameType   }

        Credentials ::= CHOICE  {
                        [0] SimpleCredentials OPTIONAL }        20

        SimpleCredentials ::= SET  {

                user  [0]  DistinguishedName,
                password [1] OCTET STRING  }


        SsUnbind ::= UNBIND
                -- There are no arguments

                                                                30


        -- BIND and UNBIND Operations between SSAs

        SsaBind  ::= BIND
                BIND ARGUMENT  SsaBindArgument

        SsaBindArgument ::= SET {
                securityAttributes [0] Credentials OPTIONAL,
                environment [1] ObjectNameType   }        40


        SsaUnbind ::= UNBIND
                -- There are no arguments

END   -- Module BindUnbind Operations
```

Figure 4.5: Storage Protocol Bind/Unbind Definitions

—— Service Operations

ServiceOperations {}
DEFINITIONS ::=
BEGIN

EXPORTS

      Add_Object, Delete_Object, Read_Object, Modify_Object,
      Ennumerate_Object, Include_Object, Remove_Object,   10
      Search_Objects, AccessControl_Handling

—— Operations

Add_Object ::= ABSTRACT_OPERATION
      ARGUMENT AddObjectArgument
      RESULT   *—— to be defined*
      ERRORS   *—— to be defined*

                                                    20
AddObjectArgument ::= **SET** {
      name [0] ObjectNameType,
      setOfAttributes [1] **SET OF** Attribute,
      qualifiers [2] ServiceControls }

Read_Object ::= ABSTRACT_OPERATION
      ARGUMENT ReadObjectArgument
      RESULT   ReadObjectResult
      ERRORS   *—— to be defined*   30

ReadObjectArgument ::= **SET** {
      name [0] ObjectNameType,
      attribRequested [1] **SET OF** AttributeType,
      qualifiers [2] ServiceControls }

ReadObjectResult ::= **SET** {
      objectName [0] ObjectNameType,
      ValueList [1] **SET OF** AttributeValue }   40

Modify_Object ::= ABSTRACT_OPERATION
      ARGUMENT ModifyObjectArgument
      RESULT   *—— to be defined*
      ERRORS   *—— to be defined*

ModifyObjectArgument ::= **SET** {
      name [0] ObjectNameType,
      ModifRequested [1] **SET OF** Modifications,   50
      qualifiers [2] ServiceControls   }

Modifications ::= **CHOICE** {
      add [0] **SET OF** AttributeValue,
      delete [1] **SET OF** AttributeValue,
      replace [2] **SET OF** ReplaceParams }

```
ReplaceParams ::= SET {
        attrib [0] AttributeType,
        old  [1] Value,
        new  [2] Vaue  }


Delete_Object ::= ABSTRACT_OPERATION
        ARGUMENT  DeleteObjectArgument
        RESULT            -- to be defined
        ERRORS            -- to be defined

DeleteObjectArgument ::= SET {
        name  [0] ObjectNameType,
        qualifiers [1] ServiceControls  }


Ennumerate_Object ::= ABSTRACT_OPERATION
        ARGUMENT  EnnumerateObjectArgument
        RESULT    EnnumerateObjectResult
        ERRORS            -- to be defined


EnnumerateObjectsArgument ::= SET {
        compoundObj [0] ObjectNameType,
        recursion [1] Recursion,
        qualifiers [2] ServiceControls  }

Recursion ::= SET {
        recursionAllowed [0] BOOLEAN,
        depth [1] INTEGER  }

EnnumerateObjectResult ::= SET OF ObjectNameType


Include_Object ::= ABSTRACT_OPERATION
        ARGUMENT  IncludeObjectArgument
        RESULT            -- to be defined
        ERRORS            -- to be defined

IncludeObjectArgument ::= SET {
        name  [0] ObjectNameType,
        compoundObj [1] ObjectNameType,
        qualifiers [2] ServiceControls  }


Remove_Object ::= ABSTRACT_OPERATION
        ARGUMENT  RemoveObjectArgument
        RESULT            -- to be defined
        ERRORS            -- to be defined

RemoveObjectArgument ::= SET {
        name  [0] ObjectNameType,
        compoundObject [1] ObjectNameType,
        qualifiers [2] ServiceControls  }
```

60

70

80

90

100

110

```
Search_Object  ::= ABSTRACT_OPERATION
                ARGUMENT  SearchObjectArgument
                RESULT    SearchObjectResult
                ERRORS         -- to be defined                    120

SearchObjectArgument  ::= SET {
                filter [0] Filter,
                recursion [1] Recursion,
                qualifiers [2] ServiceControls   }

Recursion ::= SET {
                recursionAllowed [0] BOOLEAN,
                depth [1] INTEGER   }
                                                                   130
SearchObjectResult  ::= SET OF ObjectNameType


-- Filter has been literally copied from Directory System Definition

Filter ::= CHOICE {
                item [0] FilterItem,
                and  [1] SET OF Filter,
                or   [2] SET OF Filter,
                not  [3] Filter  }                                 140

FilterItem  ::= CHOICE {
                equality [0] AttributeValueAssertion,
                substrings [1] SEQUENCE {
                        type AttributeType,
                        strings SEQUENCE OF AttributeValue },
                greaterOrEqual [2] AttributeValueAssertion,
                lessOrEqual  [3] AttributeValueAssertion,
                present  [4] AttributeType  }
                                                                   150
AttributeValueAssertion  ::= SEQUENCE
                { AttributeType, AttributeValue  }



AccessControl_Handling  ::= ABSTRACT_OPERATION
                ARGUMENT  AccessControlHandlingArgument
                RESULT          -- to be defined
                ERRORS          -- to be defined
                                                                   160
AccessControlHandlingArgument  ::= SET
                name [0] ObjectNameType,
                requestedActions [1] SET OF RequestedAction,
                appliedTo [2] AppliedTo,
                qualifiers [3] ServiceControls  }

RequestedAction  ::= CHOICE {
                add  [0] AccessControlElement,
                delete [1] AccessControlElement,
                read [2] SET OF AccessRight  }                     170

END  -- Module Service Operations
```

Figure 4.6: Storage Protocol Service Definitions

–– Distributed Service Operations

DistributedServiceOperations {}
DEFINITIONS ::=
BEGIN

EXPORTS

ChainedAdd_Object, ChainedDelete_Object, ChainedRead_Object,
ChainedModify_Object, ChainedEnnumerate_Object,   10
ChainedInclude_Object, ChainedRemove_Object,
ChainedSearch_Objects, ChainedAccessControl_Handling,
Send_Updates, Ask_for_Updates

–– Operations

ChainedAdd_Object ::= ABSTRACT_OPERATION
        ARGUMENT ChainedAddObjectArgument
        RESULT   *–– to be defined*   20
        ERRORS   *–– to be defined*

ChainedAddObjectArgument ::= **SET** {
        name [0] ObjectNameType,
        setOfAttributes [1] **SET OF** Attribute,
        qualifiers [2] ServiceControls,
        chaining [3] ChainingArguments }

                                                                                 30

ChainedRead_Object ::= ABSTRACT_OPERATION
        ARGUMENT ChainedReadObjectArgument
        RESULT   ChainedReadObjectResult
        ERRORS   *–– to be defined*

ChainedReadObjectArgument ::= **SET** {
        name [0] ObjectNameType,
        attribRequested [1] **SET OF** AttributeType,
        qualifiers [2] ServiceControls,
        chaining [3] ChainingArguments }   40

ChainedReadObjectResult ::= **SET** {
        objectName [0] ObjectNameType,
        ValueList [1] **SET OF** AttributeValue }

ChainedModify_Object ::= ABSTRACT_OPERATION
        ARGUMENT ChainedModifyObjectArgument
        RESULT   *–– to be defined*   50
        ERRORS   *–– to be defined*

ChainedModifyObjectArgument ::= **SET** {
        name [0] ObjectNameType,
        ModifRequested [1] **SET OF** Modifications,
        qualifiers [2] ServiceControls,
        chaining [3] ChainingArguments }

```
                                                                        60
ChainedDelete_Object  ::= ABSTRACT_OPERATION
        ARGUMENT  ChainedDeleteObjectArgument
        RESULT          -- to be defined
        ERRORS          -- to be defined

ChainedDeleteObjectArgument  ::= SET {
        name [0] ObjectNameType,
        qualifiers [1] ServiceControls,
        chaining   [3] ChainingArguments }
                                                                        70




ChainedEnnumerate_Object  ::= ABSTRACT_OPERATION
        ARGUMENT  ChainedEnnumerateObjectArgument
        RESULT    ChainedEnnumerateObjectResult
        ERRORS          -- to be defined

ChainedEnnumerateObjectsArgument  ::= SET {
        compoundObj [0] ObjectNameType,                                 80
        recursion [1] Recursion,
        qualifiers [2] ServiceControls,
        chaining   [3] ChainingArguments }

ChainedEnnumerateObjectResult  ::= SET OF ObjectNameType




ChainedInclude_Object  ::= ABSTRACT_OPERATION
        ARGUMENT  ChainedIncludeObjectArgument                          90
        RESULT          -- to be defined
        ERRORS          -- to be defined

ChainedIncludeObjectArgument  ::= SET {
        name [0] ObjectNameType,
        compoundObj [1] ObjectNameType,
        qualifiers [2] ServiceControls,
        chaining   [3] ChainingArguments }

                                                                        100

ChainedRemove_Object  ::= ABSTRACT_OPERATION
        ARGUMENT  ChainedRemoveObjectArgument
        RESULT          -- to be defined
        ERRORS          -- to be defined

ChainedRemoveObjectArgument  ::= SET {
        name [0] ObjectNameType,
        compoundObject [1] ObjectNameType,
        qualifiers [2] ServiceControls,                                 110
        chaining   [3] ChainingArguments }



ChainedSearch_Object  ::= ABSTRACT_OPERATION
        ARGUMENT  ChainedSearchObjectArgument
```

```
                    RESULT    ChainedSearchObjectResult
                    ERRORS         —— to be defined


ChainedSearchObjectArgument  ::= SET  {                                120
                    filter  [0] Filter,
                    recursion  [1] Recursion,
                    qualifiers  [2] ServiceControls,
                    chaining  [3] ChainingArguments  }


ChainedSearchObjectResult  ::= SET OF ObjectNameType



ChainedAccessControl_Handling  ::= ABSTRACT_OPERATION
                    ARGUMENT  ChainedAccessControlHandlingArgument     130
                    RESULT         —— to be defined
                    ERRORS         —— to be defined

ChainedAccessControlHandlingArgument  ::= SET
                    name  [0] ObjectNameType,
                    requestedActions [1] SET OF RequestedAction,
                    appliedTo [2] AppliedTo,
                    qualifiers [3] ServiceControls,
                    chaining   [3] ChainingArguments  }

                                                                      140



Send_Updates  ::= ABSTRACT_OPERATION
                    ARGUMENT  SendUpdatesArgument
                    RESULT         —— to be defined
                    ERRORS         —— to be defined

SendUpdatesArgument  ::= SET  {
                    updates [0] SET OF DERInfo,
                    qualifiers [1] ServiceControls,                    150
                    chaining  [2] ChainingArguments    }

DERInfo  ::= SET  {
                    dERIdent  [0] DERIdentType,
                    objRef  [1] ObjectNameType,
                    operation  [2] INTEGER  {
                                   Add (0),
                                   Modify (1),
                                   Delete (2)   },
                    createDateTime  [3] Time,                          160
                    entryDateTime [4] Time,
                    affectedAttributes  [4] SET OF Attribute,
                    compobjInvolved  [5] ObjectNameType OPTIONAL    }



Ask_for_Updates  ::= ABSTRACT_OPERATION
                    ARGUMENT  AskForUpdatesArgument
                    RESULT    AskForUpdatesResult
                    ERRORS         —— to be defined                    170


AskForUpdatesArgument  ::= SET  {
                    objects  [0] SET OF ObjectNameType,
                    sinceDate [1] Time,
                    Constrictions [2] Constrictions    }
```

AskForUpdatesResult ::= **SET OF** DERInfo

Constrictions ::= **SET** {
        maxDERs [0] **INTEGER**,   180
        maxTime [1] **INTEGER**,
        maxSize [2] **INTEGER**   }

END   –– *Module Distributed Service Operations*

Figure 4.7: Storage Protocol Distributed Service Definitions

```
Protocols {}
DEFINITIONS ::=
BEGIN

IMPORTS

        serviceOperations, distributedServiceOperations,
        bindUnbindOperations
                FROM UsefulDefinitions {}
                                                                        10
        APPLICATION−SERVICE−ELEMENT, APPLICATION−CONTEXT
                FROM Remote−Operations−Notation−extension {}

        BIND, UNBIND
                FROM Remote−Operations−Notation {}

        SsBind, SsUnbind, SsaBind, SsaUnbind
                FROM BindUnbindOperations bindUnbindOperations

        Add_Object, Delete_Object, Read_Object, Modify_Object,          20
        Ennumerate_Object, Include_Object, Remove_Object,
        Search_Objects, AccessControl_Handling
                FROM ServiceOperations serviceOperations

        ChainedAdd_Object, ChainedDelete_Object, ChainedRead_Object,
        ChainedModify_Object, ChainedEnnumerate_Object,
        ChainedInclude_Object, ChainedRemove_Object,
        ChainedSearch_Objects, ChainedAccessControl_Handling,
        Send_Updates, Ask_for_Updates
                FROM DistributedServiceOperations distributedServiceOperations

−− ASEs

        ssServiceASE
        APPLICATION−SERVICE−ELEMENT
        SUPPLIER PERFORMS
                { add_Object, delete_Object, read_Object, modify_Object,
                ennumerate_Object, include_Object, remove_Object,
                search_Objects, accessControl_Handling }
        ::= {serviceElement 1}                                          40

        add_Object              Add_Object  ::=  1

        delete_Object           Delete_Object ::=  2

        read_Object             Read_Object ::=  3

        modify_Object           Modify_Object ::= 4

        ennumerate_Object          Ennumerate_Object ::= 5              50

        include_Object          Include_Object ::=  6

        remove_Object           Remove_Object ::= 7

        search_Objects          Search_Objects ::= 8
```

accessControl_Handling   AccessControl_Handling ::= 9

60

ssDistributedServiceASE
  APPLICATION–SERVICE–ELEMENT
    SUPPLIER PERFORMS
        { chainedAdd_Object, chainedDelete_Object, chainedRead_Object,
        chainedModify_Object, chainedEnnumerate_Object,
        chainedInclude_Object, chainedRemove_Object,
        chainedSearch_Objects, chainedAccessControl_Handling,
        send_Updates, ask_for_Updates }
    ::= {serviceElement 2}

70

chainedAdd_Object   ChainedAdd_Object ::= 10

chainedDelete_Object   ChainedDelete_Object ::= 11

chainedRead_Object   ChainedRead_Object ::= 12

chainedModify_Object   ChainedModify_Object ::= 13

chainedEnnumerate_Object
            ChainedEnnumerate_Object ::= 14   80

chainedInclude_Object ChainedInclude_Object ::= 15

chainedRemove_Object   ChainedRemove_Object ::= 16

chainedSearch_Objects ChainedSearch_Objects ::= 17

chainedAccessControl_Handling
            ChainedAccessControl_Handling ::= 18

90

send_Updates   Send_Updates ::= 19

ask_for_Updates   Ask_for_Updates ::= 20

–– *application Contexts*

ssServiceAC
  APPLICATION–CONTEXT
    APPLICATION SERVICE ELEMENTS {aCSE}   100
        BIND ssBind
        UNBIND ssUnbind
        REMOTE OPERATIONS {rOSE}
        INITIATOR **CONSUMER OF** {ssServiceASE}
        ABSTRACT SYNTAXES {ssAS}
    ::= {applicationContext 1}

ssBind   SsBind   ::= **NULL**

ssUnbind   SsUnbind ::= **NULL**   110

ssDistributedServiceAC
  APPLICATION–CONTEXT
    APPLICATION SERVICE ELEMENTS {aCSE}
        BIND ssaBind
        UNBIND ssaUnbind

REMOTE OPERATIONS {rOSE}
INITIATOR **CONSUMER OF** {ssDistributedServiceASE}
ABSTRACT SYNTAXES {ssAS}
::= {applicationContext 2}   120

ssaBind   SsaBind   ::= **NULL**

ssaUnbind   SsaUnbind ::= **NULL**

ssAS   ::=   {abstractSyntax 1}

END

Figure 4.8: Storage Protocol Definitions

Chapter 5

The Distribution Service

Authors: Lanceros Andres
            Juan Saras

The distribution service provides the means for the distribution of Information Objects (IOs) inside a group or between groups. These IOs may consist of:

- IP messages.

- Forms.

- Files, File Access Data Units (FADUs from FTAM), set of files.

- Documents, ODA documents, SGML documents, sets of documents.

- EDI formats.

- Communication contexts (e.g. conversations).

- Conferences.

- Data base event records (see Archive Services chapter).

and others.

Their distribution requires the kind of functionality which is capable of transferring IOs in a transparent manner and cooperating with other services. This functionality is defined by the Distribution Service outlined in this chapter.

5.1   Functional Overview

The Distribution Service (DiS) is one of the Basic services used by the Group Communication Coordination Service (see chapter three).

The DiS may need to interact with other services (belonging to the Basic set or not) in order to perform its distribution function. For example, a Group Communication Agent (GCA), acting on behalf a group, could request the DiS to distribute an IO, passing it by value to the DiS. Alternatively a GCA might request the DiS to distribute an IO by reference (providing some kind of object description, instead of the IO itself). In this case the DiS, acting on behalf of the GCA, might access a Document Filing and Retrieval System to retrieve a set of documents; the documents subsequently being distributed inside the group co-ordinated by that GCA (possibly in cooperation with other GCAs). The actual distribution could also be performed by storing the set of documents in an ASA where they would be accessed subsequently by group members.

The Distribution Service is specialized to the distribution of IOs inside and between groups. It is not intended to duplicate the kinds of replication facilities between Storage Agents described in the previous chapter. Furthermore, the Distribution Service is not intended to act as an Archive Service. It is assumed to have only transient storage capacity i.e., every IO submitted to the DiS will pass out through a distribution port after some period of time.

5.2   Architectural Overview

The Distribution System may be viewed as being composed of two entities:

- A Distribution User Agent (DiUA) which acts on behalf a user to access the distribution system and distribute submitted Information Objects. The submitted objects could either be references to IOs stored in some system (e.g. the Archive System or the Document Filing and Retrieval System), or the IOs themselves.

- A Distribution Service Agent (DiSA) which performs the distribution either by itself (e.g. by directly providing a copy of the IO to each group member), or with the help of another system (e.g. by storing the IO within an ASA to be accessed later on), with or without the collaboration of other DiSAs. When an IO is provided by reference a Third Party Transfer may be started (e.g. to transfer a file from a File Store, as it is defined in FTAM, to an ASA).

Between the above entities, two protocols may be defined:

- A Distribution Access Protocol (DiAP) between a DiUA and a DiSA.

- A Distribution Service Protocol (DiSP) between DiSAs.

Both protocols could use as a transfer medium either a store-and-forward solution (e.g. MHS-84) or a connection oriented solution (e.g. ROS). Some transfer media may have their own distribution facilities (e.g. MHS-88) that could facilitate the distribution task.

Distribution List based Services

Several architectural possibilities for the realization of a Distribution Service have been outlined above. However, in what follows, we shall concentrate on the use of Distribution Lists to implement the Service. These provide a relatively simple means by which some of issues raised above can be explored further. (This decision also implies that Information Objects will be always passed by value.)

The present AMIGO MHS+ piloting work has provided a specification for the distribution of IP-Messages via Distribution Lists in the MHS 1984 User Agent Layer (see the DL piloting section in chapter seven). In this simple case, the correspondence between the above entities and protocols could be described by the following:

- DiUA = UA

- DiSA = GA (in fact, a UA with some kind of added functionality).

- DiAP = P2

- DiSP = P2+ (in fact, P2 plus a few new rules to avoid duplication and loops).

As has already been pointed out, if a transfer media has its own distribution facilities (as is the case with MHS-88), the DiS could use them for its own purposes. In this case, a similar mapping is possible to that defined above:

- DiUA = UA

- DiSA = MTA (in fact, an MTA following the 88 version of MHS)

- DiAP = P3

- DiSP = P1 (in fact, the 88 version of the P1 protocol)

In summary the distribution of IOs may be performed using DLs expanded within a special UA or expanded inside the MTS. From a Group Communication viewpoint both kinds of distribution have similar characteristics (in terms of associated information, management, etc.).

However, in general, UA list expansion offers the opportunity to perform more "intelligent actions" (i.e., within the more flexible and end-user accessible UA environment). Examples of such intelligent actions might be the automatic grouping of all responses to a message or the automatic counting of votes in decision making.

On the other hand, expansion inside the MTS allows the use of the distribution service by any kind of message the MTS is able to transfer (i.e. not only IPM messages). In addition, as all modifications are made on the P1 envelope, every final recipient receives the same content copy without any UA enforced changes.

5.3   Distributors

A Distributor should define all information needed to perform the distribution of an IO inside or between groups. The minimum required information may be the following:

Initiators A set of Distinguished Directory Names (DDNs) that can invoke distribution through this distributor.

Destinations A set of DDNs to which a distributor should distribute IOs given to it by the Initiator.

Any distributor should have a Distinguished Directory Name (DDN) and the above information may be kept inside the Directory Service. It is clear that the concept of distributor, in the simple case described here, can be mapped onto the concept of a Distribution List.

A mapping from the above terms to those common in distribution lists is the following:

- initiators = submitting members

- destinations = receiving members

In addition to the minimum information indicated above, additional roles may be required to fully characterise a Distribution List. Examples could include the auditor, the person responsible for maintaining the DL, or the DL owner. The DL piloting section of chapter seven discusses this issue in more depth.

In principle, the destination indicated by a distributor might be just an ASA where group members could access the distributed IOs. However, this would mean that the relation between Groups and Distributors would have to have been well defined and it is to this subject that we now turn.

5.4   Groups versus Distributors

We assume that the distribution of contributions inside a group or between groups will be performed by the DiS through a set of linked DLs (located inside UAs or within the MTS). Therefore, the first requirement is to know which set of DLs are associated with a given group. Since both the group and the DL entities will be Directory entries, it seems reasonable that the DLs or the group itself might have some attribute showing this relationship. Therefore, two solutions might solve our problem:

- A DL could have an optional attribute to store the Directory name of the group to which it belongs (let us call it a "group attribute").

- The group definition inside the Directory may have an attribute which explicitly records the set of DLs used to distribute contributions. (This attribute, of course, should be recurrent.)

The first solution is a distributed one. The information about how many DLs perform the distribution is dispersed throughout all the DLs belonging to the group. This information would be maintained by the Directory. To know which set of DLs are associated with a given group, a user directory agent would only have to ask about all DLs satisfying the condition "group attribute = required group name".

The second solution is a centralised one. All the information is contained inside the group entry. Its maintenance would have to be explicitly performed by the group administrator (e.g. a GCA).

Of the two solutions presented above, the first seems more preferable and this will be assumed in the following sections.

5.5   Distribution of contributions

When a group member wants to distribute or input a contribution to a group, he/she should only need to provide to the GCS the group name to which the contribution should be sent. The GCS will then use the DiS to actually perform the distribution.

According to the previous discussion, to distribute a message in a group means to choose one DL from among a set of potential DLs (those associated with the group). The chosen DL should be an "entry point" to the group for the submitting user. An entry point could be defined as a DL satisfying the following condition: "every contribution addressed to an entry point (DL) will be distributed to all members of the group". (It is assumed that, in general, not all DLs associated with a group will be entry points). Also, the entry point for a group may vary with the group member for technical reasons (e.g. the chosen DL should be the closest to the GCA requesting distribution) or administrative reasons.

Once a contribution is sent to a group entry point DL, the List will be recursively expanded and a copy will be delivered to each final destination. As already mentioned, some of these may be ASAs where other group members can access the distributed contribution. Of course, it must be the task of the GCS to ensure that the set of final recipients (direct recipients and ASAs) is consistent.

5.6   Marking of contributions

It is necessary, by marking (or stamping) contributions with a group name, to allow recipients to determine that a received item has passed through a distributor. For example, a UA acting as a direct recipient of a DL might like to be able to identify messages as coming from a given group. (A similar effect could be obtained by accessing the Directory to discover to which group a DL belongs. Nevertheless, although this information might be accessible to an ASA, this may not be true generally for more common agents such as normal recipient UAs.)

The marking could be performed in two places:

- Inside a GCA *before* distributing the contribution through the appropriate set of DLs.

- When the DLs are expanded, as an additional operation to be performed.

Now, consider the case where two established groups are going to cooperate. One way of linking them would be to add the entry points of each group as recipients of some DL of the other group. Then, when a contribution crosses the border between both groups, it will have been previously marked (by either of the previous marking strategies) as belonging to one of these two groups. However, the contribution will also belong subsequently to the second group and will be marked as such. This would give two 'owners'. The solution would be for each group's entry point DL, to mark the contribution as belonging to its group *only* if it had not been previously marked. This solution suggests that the second of the above marking alternatives be adopted.

5.7   Service Operations

The basic operation performed by the Distribution Service is:

- Distribute (Initiator, Object_Descriptor, Distributor)

where "Initiator" is the DDN of the entity that makes the distribution request (most frequently a GCA), "Object_Descriptor" is any combination of the IOs identified as potential objects to be distributed (or references to them) and, finally, "Distributor" is the DDN of a distributor as defined in the previous section.

Its semantic is a request by the initiator entity for a set of objects identified by the object descriptor to be distributed by means (or through) the indicated distributor.

Finally, another mapping to the DL environment is defined here. Therefore, the Distribute operation is reduced to the sending by the "initiator" of the IOs defined by the "object descriptor" (passed by value) to the distribution list (the "distributor") that actually will perform the distribution.

Summary

This chapter has outlined how the abstract Basic Distribution Service described in chapter three might be realized. The AMIGO MHS+ group piloted aspects of this service during the project and the results are described in chapter seven.

Chapter 6

Use of the Directory Service for Group Communication Support

Authors: Karl-Heinz Weiss
Contributors: Oliver Wenzel
                Michael Tschichholz

6.1   Introduction

This chapter examines how the ideas underlying group distribution agents and distribution lists can be extended to form the basis of the more general concept of a 'communication group environment'.

Group communication can take several forms. One form is a one-to-all communication which may be realised by the use of distribution lists. Another form is an all-to-one communication, which might be represented by a voting procedure. All-to-all communication can be considered as a combination of previous forms, e.g. conferencing. However, for our purposes, it is more convenient to characterize group communication by the environment in which group members are found rather than by special types of communication. That is, in mapping communication groups and environments onto the Directory Service (DS) there will be no requirement to have distinct structures for special communication patterns. Instead, special semantics of communication patterns within can be re-introduced by assigning suitable attributes to the entry of a communication group environment.

6.1.1   Representing communication group environments in the DS

The DS can be used as a mechanism to provide globally available information about communication groups and their environment, so that someone who needs information about a particular communication group can use the Directory System operations to obtain it.

The organisational relations in a group communication environment can be structured as a tree, similar to the structuring of the ISO/CCITT directory information tree (DIT). The term ISO/CCITT directory information tree will serve to identify the set of possible tree structures which can be created using the object-class definitions provided by the standard [ISO87f]. The possible hierarchical relations are described by DIT structure definitions, which define explicitly for each object class it's possible subordinates in the tree structure. This ISO/CCITT tree is oriented to a specific organisational structure, i.e. that the world (the root) consists of countries or international organisations or that an organisation consists of organisational units, -persons, -roles, etc.

As an enhancement to those predefined object-classes and structure definitions it is possible for an administrative authority (authority of a naming context) to define a set of new object-class and DIT-structure definitions.

There are two possible ways to integrate the communication group environment entry in the DIT:

1. Each communication group environment is placed as a subordinate to the entry to which it has a relation, e.g. a communication group within the environment HMI will be a subordinate entry of the entry for the Hahn-Meitner-Institut organisation (HMI).

2. Different levels of communication group environments, such as international, national etc., can be grouped as sub-trees of the particular entries for "international communication group environments" or "national communication group environments".

Because both ways have certain advantages for particular group structures, they should be used and both need to be supported.

In the first approach national communication group environments or organisation-internal communication group environments could be defined as subordinates of already existing entries in the DIT.

The second approach is useful for international communication group environments which are not related to a particular organisation or country. It permits several "sub-roots" for international communication group environment subtrees, which will minimize the number of direct root subordinates. Because of the nature of the internal organisation of the Directory System, DSA's which hold entries directly subordinate to the root require additional administration. Therefore a special object class "international communication group environment" which will be valid as a root subordinate must exist. This will cause a modification of the DIT structure rule for the object class "root", which currently allows using the object classes "country" and "organisation" as root-subordinates.

Because of the possibly large number of communication group environments and the possible sub-structuring, it could be an advantage to have special DSA's as administrative authorities for particular communication group environment sub-trees of the DIT. If those DSA's reside at the same location as other group-communication services, this mechanism will result in easy access to the communication group environment information required by those services. This makes the first approach preferable, because of the possibility of better separation i.e. using special DSA's for group support.

6.1.2 Representing communication group members in the DS

There are two different ways of mapping the relationship between a group member and the communication group environment onto the DS structure:

- each group member is described by a value of a special attribute in the entry for the communication group environment.

- each group member is represented by an entry of a special object-class as a subordinate of the entry for the communication group environment.

Representation by Attributes

The use of attributes to represent the members of a communication group environment has the advantages that the description of the environment and the description of the members are contained in the same entry and so can be added or deleted as a whole. Furthermore there are fewer entries which have to be administered by the DS.

But this approach also has several important disadvantages. If more attributes need to be added to the description of each group member (e.g. special group permission, roles into the group), this might be impossible with this approach. Moreover, access controls for each particular value of the attribute describing the group members are required because each member should have access to the parameters describing his/her group-membership. This prevents a member from being able to remove himself from the list of group members, otherwise he will have to have access rights to modify the whole set of attributes.

Representation by Entries

This approach permits a more flexible assignment of access controls to group member descriptions. A particular group member can have access rights to modify his own entry, e.g. modifying his personal group membership parameters, but can be prevented from modifying entries of other group members. Access rights to add or remove a subordinate entry of the group entry can be assigned to persons, groups or environments to allow self-joining to the group. Another advantage is that a group-membership entry of a person can be shared for different communication group memberships using the alias mechanism of the DS.

Because of the increased number of directory objects which have to be administered using the entry-approach the problem may arise that the DS will be overloaded. This problem can partially be avoided by storing group communication environment subtrees at special group communication DSA's as mentioned above.

Because of its advantages, the entry approach seems to be more preferable and therefore will be used in the following.

6.1.3   The group member entry

There are again two possible ways to represent an entry of a communication group member:

- using an alias reference to the primary entry of the member person

- using an entry of a special object class "group member"

Using Alias Entries

If the DS alias mechanism is used, attributes related to membership must be stored in that primary entry. Example: if the primary entry for a person is (Country=DE; Organisation=GMD; OrganisationalUnit=Berlin; OrganisationalUnit=FOKUS; OrganisationalPerson=Berthold Butscher), an alias entry for the primary entry can be used to represent the membership of this person in the communication group environment (InternCommGroupEnvironment=AMIGO CommGroupEnvironment=MHS+). Because a person can be a member of multiple communication group environments, there must either exist distinct sets of attributes for each group membership or a set of standard attributes which will be valid for each group membership. Further, only people who are already registered by the DS can be referred to using alias entries.

Using Specific Entries

The use of an entry of a specific object class provides a mechanism for specifying different attributes for each group membership of a particular person. By aliasing an entry of that class, sharing membership entries over different group memberships is also possible. Under the assumption, that communication environment subtrees are stored at special DSA's in conjunction with group communication services, the use of special entries leads to a fast and cheap access of the membership data.

Beside the set of attributes specifying group membership conditions there must be attributes related to the description of the particular person. This could be done by including all necessary personal attributes (e.g. a subset of the attributes of the primary entry of the person) in the membership entry, but this will result in a lot of redundant data stored in the membership entry.

Another approach is to include an attribute containing the distinguished name of the primary entry which provides a way of accessing personal data. If the personal data is accessed, this can be done by issuing a subsequent read-request to the DS using the distinguished name contained in the membership entry. Another significant advantage is the possibility of providing better security for personal data. Because of the separation of membership data and personal data it is possible to have different access controls thereby providing more selective access to personal data. By this means a group communication service can be permitted to have access to the membership entry but is denied access to personal data of the member.

Because of the possibility of environment specific attributes and the separation of group communication environment related data from other DS entries, the use of a membership entry is preferred. The personal data are referenced by a distinguished name because of less redundancy and better data protection capability.

6.1.4   Modelling Environment Relationships

A Communication Group Environment can be a member of other Communication Group Environments. No implicit membership relationships exist between these environments and Members of the environments. That means, that required membership relations between these environments and members of the "sub-environment" must be defined explicitly. There are three distinct forms of such membership:

- A Communication Group Environment is a real subgroup of another one. The entry for the subgroup is represented as a subordinate entry of the parent environment. Environment members of that "sub-environment" belong to the parent environment.

- A Communication Group Environment is a guest member of another environment. If an already existing Communication Group Environment becomes a member of another environment, the distinguished name of its already existing entry is used. This permits the use of all attributes of the member entry to be used to define the guest membership relation of the environment.

- An already existing Communication Group Environment becomes a "sub-environment" of another environment. This can be modelled using an alias entry for the already existing environment entry as a subordinate of the entry of the environment which is to be joined. The attributes related to the Group Member entries of the attached sub-environment are valid for the membership in the parent entry.

6.1.5   Definition of the DS Object-Class "International Communication Group Environments"

The following set of information can be contained in the InternComm-GroupEnvironments object class:

Name

This attribute contain the relative distinguished name (on the country level) of a set of Communication Group Environments which are clustered to one of a few (!) distinct International Communication Group Environments (e.g. english-language-conferences). To avoid administrative problems within the DS only a limited number of these entries would be allowed (e.g. < 100).

Group description

A textual description of the common "things", which leads to the clustering of all contained Communication Group Environments.

6.1.6   Definition of the DS Object-Class "Communication Group Environment"

The following is a list of possible properties of a communication group environment description and their mapping onto directory attribute types and attribute values.

Name

This attribute contains the relative distinguished name of the entry which is part of the whole distinguished name for the communication group environment.

Environment description

A textual description of the purpose of the environment. This information can be useful for a prospective new member. The text should also contain information about the joining procedure for this environment (e.g. send a message to the moderator/administrator, or self-add via a directory add-object operation in the case of environments with open groups).

Current Topics

A textual description of the current communication topics in that environment.

Public Documents

A set of selected documents could describe the interface to the outside world. Thus it is possible to make a document reference globally public. An expanded approach will be described in section 6.2.2.

Roles

In a communication group environment there will exist a set of rules that can be played by different persons. This relation between persons and roles should be included in the communication group environment description. One solution is to include for each possible role a special attribute type which will limit the spectrum of roles to those which have be predefined. Another way is to describe role assignment using a textual description for a sequence of assignments in a single attribute.

Rules

Communication in an environment will be guided by certain rules which define the communication patterns used in the communication processes of the groups within these environment. Because this is a very important characteristic of the environment it should be represented in the entry, allowing a group communication agent to access those rules. A representation of the rules themselves in a directory attribute will not be practicable because of the complexity that such rule descriptions can have. Moreover, the way of rules are described can differ between the several group communication agents which have to apply them. Therefore it seems to be more practicable to have a set of predefined categories for communication rules. The use of particular patterns in a specific environment can then be represented by an attribute containing a sequence of categories for Communication rules.

O/R Address

The O/R addresses of the Group Communication Agents for this environment.

Archive Agent address

One or more Archive Agent Addresses which are designated to store contributions related to the group. An ASN.1 description for the object-class CommGroupEnvironment, related attributes and the DIT structure rules is given in section 6.5. The description technique is similar to that one used in the ISO/CCITT DS standards [ISO87c] using the ASN.1 macros defined in [ISO87d] and the attribute types and syntax defined in [ISO87e].

6.1.7　Definition of the DS Object-Class "Group Member"

This can be considered as a proposal for the set of information contained in a Group Member entry.

O/R Address

An O/R address that can be used by a group communication agent to deliver parts of the messages of the group to the particular group member.

DS name

The distinguished name for the primary entry of the group member in the DS. This primary entry may contain further personal data about the group member. If the member is not registered elsewhere in the DS, this attribute can be empty. This can also be the DS name of another Communication Group Environment.

Roles

If the member plays a special role in the group communication environment (e.g. administrator, moderator, editor, chairman) this can be represented by this attribute.

Mode

This attribute can be used by a group communication service to select the mode by which messages will be delivered to the member. Examples for modes are auto-deliver (member receives every message automatically), deliver-on-demand (member demands service to deliver a new message) or auto-summary (member receives only a summary of the subjects of messages automatically). This is only one possibility for the use of the mode attribute. The concrete meaning of the attribute is a matter of the used communication service (for further study).

Default Archive-Agents or Group-Communication-Agents

In the case that a distributed communication service is used, a default service entity can be selected which will be used by the service to handle special actions related to that member (e.g. storage of contributions submitted by the member).

In addition there will be many more attributes which could be useful for a particular communication service. However, it will be a local matter for those services to enhance the set of attributes.

An ASN.1 description for the object-class GroupMember and related attributes is given in section 6.5.

6.2 DS support for the Archive System

6.2.1 Environment Archives

In the proposed object class definition for communication group environments we have defined an attribute type which contains "address(es)" for environment specific archives. This concept enables a dynamic relation between environments and (distributed) archives e.g. for optimal load balancing and reduced communication overhead.

6.2.2 Relation between Group and Group-specific public documents

To enable public access to environment-specific documents the following scheme provides three possible ways of achieving the necessary functionality:

- An informal description of public documents could be realized by the use of a "publicDocument" attribute as described in section 6.1.6.

- To allow functional use of public document information by service entities and to be able to retrieve document specific attributes by the DS, an informal description is not sufficient. To support these requirements, a document specific DS entry is needed. These entries will be subordinates of communication group entries. However, this mechanism could not replace advanced retrieval capabilities provided by an archive system itself, it is just an additional feature. A misuse of this feature might be dangerous because of the possible overload of the DS and must therefore be prevented.

- A third possibility would be the storage of all group specific document references in the DS using the second approach. Contrary to the above approaches, where the archive system provides the retrieval functionality, the third approach would enable the use of DS as retrieval system for group specific document references. But this approach is in contradiction to basic principles and therefore a misuse of the DS. Because of the large number of documents this would result in very large subtrees which would have to be managed by the DS. Furthermore, the nature of document storage implies very high update rates which contradicts the assumption of a low update rate for the directory information base. For these reasons we conclude that one should only store selected document references within the directory, to provide a global visibility to the outside world.

6.2.3   Definition of the DS Object Class "Public Document"

The following provides an outline definition and this needs further expansion.

An entry for a public document description can consist of the following set of attributes:

Document Id

A unique id which can be used to identify the document which belongs to the document-description entry. This id can be used to request the document itself at a document store.

Archive Agent Address

A reference to the document archive agent, which store a copy of the document and can be used to receive a document.

Subject

Subject of the document.

Author(s)

Author(s) of the document.

The set of all possible attributes of a document description depends on the different document characteristics which can be used in search/list requests.

A further enhancement could be the use of special entries, such as cabinet, folder, etc., for structuring sets of documents which are integrated into a group communication environment subtree.

6.3 Evaluation Summary of DS Use

The currently proposed DS standard allows only limited and artificial Group Communication usage due to the lack of suitable object classes for group communication environments. However, it has been shown that the definition of the object class 'communication group environment' leads to a powerful support for group communication. Therefore DIT-rules for the object classes country, organisation and organisational unit should be expanded to allow the object class 'communication group environment' as a subordinate.

International communication group environments (e.g. MHSnews, english-language conference, unix-user, etc.) may be distributed over several countries and may not be related to already defined object classes. These groups cannot be subordinates of existing entries of the DIT. To support such groups, another object class (international communication group environments) as root subordinate is required. This avoids thousands of entries on the country level, which would lead to administrative DS problems.

The possible ways in which the DS can handle extensive usage by the 'Group-Communication-System' without a significant reduction in performance is an important topic. Therefore this should be evaluated in the near future, probably by early prototype versions of the DS. Because the management of replicated DIB data will be an important factor, this evaluation depends on the availability of the further developments of the DS (study period 1988-92), which will incorporate replication mechanisms in the DS.

6.4 Examples of the AMIGO use of DS

The following examples are intended to illustrate how to use the DS in the group communication context for conferencing. Many other uses are also possible.

6.4.1 Current DIT structure related environments

We can use the already defined DS object classes as subordinates for object class CommGroupEnvironment, if groups are related to countries, organisations, organisational units, etc. This is shown in Figures 6.1 and 6.2.

6.4.2   Expanded DIT structure for international communication group environments

Because there are communication groups which cannot be mapped onto the current DIT structure, the object class international communication group environments should be added as root subordinate. This allows a flexible definition of potential possible environments as shown if Figure 6.3.

6.5   Directory Object Class and Attribute definitions

The Object Classes and Attributes which have to be integrated into the already defined Object Classes and Attributes of the DS are given using the ASN.1 notation. This notation uses macros defined in [ISO87d] and a set of standard attribute types which are defined in [ISO87e].

The following is not a complete definition. Instead it is intended to provide an overview of how such a definition might appear.

6.5.1   Object Class and Attributes for InternCommGroupEnvironments

```
internCommGroupEnvironment OBJECT-CLASS
        SUBCLASS OF tOP
        MUST CONTAIN {
                commonName,
        }
        MAY CONTAIN {
                description
        }
        ::= {objectClass ??}
```

Attribute Types

```
description ATTRIBUTE-TYPE
        ATTRIBUTE-SYNTAX caseIgnoreString
        MULTIPLE VALUE
        ::= {attributeType 10}
```

Rule definition

The Rule Definition describes the possible positions in the DIT for entries of the object class InternCommGroupEnvironment.

```
RULE-FOR internCommGroupEnvironment
     SUBORDINATES {
     commGroupEnvironment { NAMES commonName }
     }
     ::= {dITStructure ??}
```

The DIT-rules for the object class root should be changed to allow intern-CommGroupEnvironment as subordinate.

6.5.2 Object Class and Attributes for CommGroupEnvironment

```
commGroupEnvironment OBJECT-CLASS
     SUBCLASS OF tOP
     MUST CONTAIN {
          commonName
     }
     MAY CONTAIN {
          EnvironmentDescription,
          SET OF ASA,             -- Archive-System-Agents
          SET OF GCSA,            -- GCSAs
          currentEnvironmentTopics,
          publicDocuments
          roles,
          rules
     }
     ::= {objectClass ??}
```

Attribute Types

```
EnvironmentDescription ATTRIBUTE-TYPE
     ATTRIBUTE-SYNTAX caseIgnoreString
     MULTIPLE VALUE
     ::= {attributeType ??}

ASA ATTRIBUTE-TYPE                      -- Archive-System-Agents
     ATTRIBUTE-SYNTAX ApplicationEntity
     MULTIPLE VALUE
     ::= {attributeType ??}

GCSA ATTRIBUTE-TYPE                     -- GCSAs
     ATTRIBUTE-SYNTAX ApplicationEntity
     MULTIPLE VALUE
     ::= {attributeType ??}


currentEnvironmentTopics ATTRIBUTE-TYPE
```

```
      ATTRIBUTE-SYNTAX caseIgnoreString
      MULTIPLE VALUE
      ::= {attributeType ??}


publicDocuments ATTRIBUTE-TYPE
      ATTRIBUTE-SYNTAX caseIgnoreString
      MULTIPLE VALUE
      ::= {attributeType ??}


roles ATTRIBUTE-TYPE
      ATTRIBUTE-SYNTAX caseIgnoreString
      MULTIPLE VALUE
      ::= {attributeType ??}


rules ATTRIBUTE-TYPE
      ATTRIBUTE-SYNTAX integerSyntax
      MULTIPLE VALUE
      ::= {attributeType ??}
```

Rule definition

The Rule Definition describes the possible positions in the DIT for entries
of the object class CommGroupEnvironment.

```
RULE-FOR commGroupEnvironment
      SUBORDINATES {
      commGroupEnvironment { NAMES commonName }
      groupMember { NAMES commonName }
      publicDocument { NAMES commonName }
      }
      ::= {dITStructure ??}
```

The DIT-rules for the object classes root, country and organisation should
be changed to allow commGroupEnvironment as subordinates.

6.5.3   Object Class and Attributes for GroupMember

```
groupMember OBJECT-CLASS
      SUBCLASS OF tOP
      MUST CONTAIN {
            commonName }
      MAY CONTAIN {
            oRAddress,
            dsName,
            mode,
            role,
```

```
                    defEntities }
          ::= {objectClass ??}
```

Attribute Types

```
    dsName ATTRIBUTE-TYPE
          ATTRIBUTE-SYNTAX distinguishedNameSyntax
          MULTIPLE VALUE
          ::= {attributeType ??}

    mode ATTRIBUTE-TYPE
          ATTRIBUTE-SYNTAX caseIgnoreStringSyntax
          MULTIPLE VALUE
          ::= {attributeType ??}

    role ATTRIBUTE-TYPE
          ATTRIBUTE-SYNTAX caseIgnoreStringSyntax
          MULTIPLE VALUE
          ::= {attributeType ??}

    defEntities ATTRIBUTE-TYPE
          ATTRIBUTE-SYNTAX caseIgnoreStringSyntax
          MULTIPLE VALUE
          ::= {attributeType ??}
```

6.5.4 Object Class and Attributes for publicDocument

```
    publicDocument OBJECT-CLASS
          SUBCLASS OF tOP
          MUST CONTAIN {
                commonName,
                documentId,
                asaAddress }    -- Archive-System-Agent Address
          MAY CONTAIN {
                subject,
                SET OF Author,
                ...     }
          ::= {objectClass ??}
```

Attribute Types

```
    documentId ATTRIBUTE-TYPE
          ATTRIBUTE-SYNTAX documentIdSyntax
          MULTIPLE VALUE
          ::= {attributeType ??}
```

```
asaAddress ATTRIBUTE-TYPE
    ATTRIBUTE-SYNTAX asaAddressSyntax
    MULTIPLE VALUE
    ::= {attributeType ??}

subject ATTRIBUTE-TYPE
    ATTRIBUTE-SYNTAX caseIgnoreStringSyntax
    MULTIPLE VALUE
    ::= {attributeType ??}

author ATTRIBUTE-TYPE
    ATTRIBUTE-SYNTAX distinguishedNameSyntax
    MULTIPLE VALUE
    ::= {attributeType ??}
```

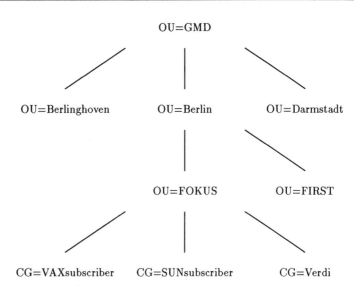

Figure 6.1: Object Classes related to standard objects

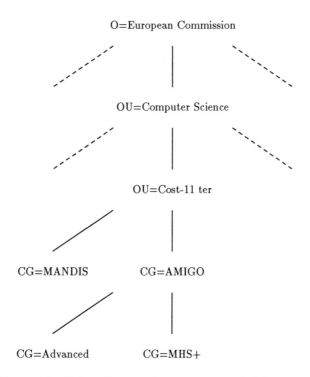

Figure 6.2: Object Classes related to standard objects

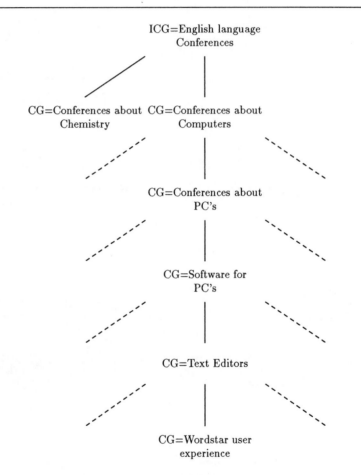

Figure 6.3: Expanded DIT structure

Chapter 7

Piloting

Authors:

GMD section: Bernd Wagner
Manfred Bogen

Madrid Section: Miguel Nuñez
Justo Gallardo

Nottingham Section: Steve Benford
Julian Onions

QZ Section: Jacob Palme

7.1 Introduction

This chapter describes prototyping activities which occurred as part of the AMIGO MHS+ project. Prototyping provided a testbed for various AMIGO ideas and therefore generated valuable feedback for the project as a whole.

Prototyping occurred in three main areas:

- Implementation of a pilot distribution list protocol.

- Implementation of a pilot directory service supporting the distribution list protocol.

- Implementation of a pilot multi-user message store.

These activities involved the Nottingham, GMD, QZ and Madrid partners as shown by the following table.

partner	distribution	directory	message store
Nottingham	yes	yes	no
GMD	yes	no	no
QZ	no	yes	no
Madrid	no	no	yes

Due to the frequent interworking of local systems, this chapter describes piloting activities as they occurred at each site as opposed to grouping them by activity. Furthermore, distribution list piloting utilised a specific AMIGO protocol, supported by 1984 X.400 and different from that described in chapter 5. This chapter provides an outline of this protocol.

The remainder of this chapter is structured in the following way:

- Overview of the AMIGO distribution list protocol

- Description of Nottingham piloting

- Description of GMD piloting

- Description of QZ piloting

- Description of Madrid piloting

7.2   Overview of the AMIGO distribution list protocol

This section provides an overview of the AMIGO distribution list protocol providing distribution list functionality within X.400 message handling systems. A full description of this protocol is given in [BO87].

The AMIGO protocol uses X.400 (1984) as a basis for distribution list handling. In particular, it builds distribution list functionality on top of the X.400 P2 layer by introducing the concept of a P2 entity called a *Group Agent* which is responsible for expanding messages sent to a list, checking users permissions to send to the list and detecting and preventing loops occurring via nested lists.

The choice of a P2 based protocol might seem strange given the inclusion of distribution list facilities in the P1 layer of 1988 X.400. It is worth clarifying the reasons for making it.

- It is easier for a group of implementors such as the AMIGO members to create P2 entities than to modify the Message Transfer layer.

- A P2 entity can be extended to include advanced distribution list facilities such as moderated and digested lists. This is more difficult in P1.

- Where piloting experiments are being conducted using an organisation's existing mail service, the P2 solution will introduce far less disruption into the message service than would a P1 solution.

The following sections outline the major features of the AMIGO protocol.

7.2.1 The abstract distribution list

The AMIGO protocol defines distribution lists in an abstract way as a sequence of properties which can be used to implement a variety of flavours of distribution list. Examples of these types are *Moderated Lists* where all submissions to the list are vetted by a moderator before being distributed, *Open lists* where anyone may add themselves to the distribution list or *Closed lists* where membership is tightly controlled. The AMIGO protocol identifies the following properties of distribution lists:

- Name

- Description

- Charging Algorithm/Joining Procedure

- Submitting Members

- Receiving Members

- Auditor

- Moderator

- Administrator

- ReplyTo

The *name* of the distribution list is the O/R-name of the P2 entity which receives all contributions to the list. The *description, charging algorithm* and *joining procedure* are human readable descriptions of aspects of the list.

The *submitting members* and *receiving members* properties contain the names of entities who may send to the list and receive from the list respectively. The separation of these properties allows great flexibility in defining lists: for example, we can define a *team* where all receiving members can also submit to the list or a *newspaper* where a small group of people submit and many receive.

The *moderator* property names the entity who receives all unauthorised submissions to the list (i.e. where the sender is not a submitting member). The moderating function is an important part of distribution list operation and in some cases we may wish to create a *moderated list* with no submitting members so that the moderator vets all submissions.

The *Auditor* of the list is the entity which receives delivery reports associated with the list and the *Replyto* property lists the default value of the P2 *inreplyto* field of messages from the list. Finally the *administrator* of the list is responsible for maintaining access controls which permit users to alter the properties of the list.

The protocol also defines operations to read and modify list properties, maintain the list membership and create and delete lists. These operations are listed below.

Retrieve Member	Return values of submitting/receiving members.
Verify Member	Determine if entity is sub/rec member.
Describe List	Returns a set of requested properties of a list.
Add Member	Adds a new submitting or receiving member.
Delete Member	Deletes an old submitting or receiving member.
Modify Property	Modifies one of the properties of the list.

The combination of properties and operations described above specifies the abstract distribution list.

Distribution lists are allowed to be nested to reflect groupings and associations of human beings. Nesting means that one list name may be included as a receiving member of another and this may occur arbitrarily allowing looped lists (lists which contain each other). The protocol is responsible for detecting and preventing loops as messages are distributed from list to list. Loop control is an important part of the AMIGO protocol and is described more fully later.

7.2.2   The group agent

Implementation of the AMIGO distribution list protocol is divided between two main entities:

The *Group Agent* and the *Directory Service*.

The Group Agent is a P2 entity responsible for distributing messages sent to the list. It can be viewed as a modified mail box which receives messages, operates on them and sends them back to the message transfer system for delivery. The group agent must check a user's permissions to send to a list, route unauthorised submissions to the list moderator and send copies of authorised messages to the recipients of the list. The name of the group agent is the O/R-name of the list and thus anyone wishing to submit to the list must send their submission to the group agent for expansion.

The Directory is the database which stores and maintains the properties of the list. It supports the group agent by supplying the information required for expansion and is responsible for maintaining the membership and other list properties. The protocol does not assume the existence of a distributed directory. The directory service is local to the group agent and the protocol between them is a local matter. The protocol does describe a method of providing remote access to local directories by encoding directory requests and results as body parts of IP-Messages. This is intended as a short term solution until a distributed directory service based on the *Remote Operations Service* is available.

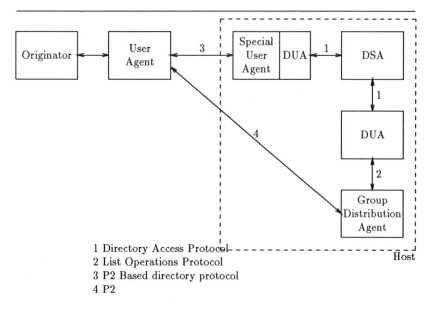

1 Directory Access Protocol
2 List Operations Protocol
3 P2 Based directory protocol
4 P2

Figure 7.1: Functional Model of the Distribution List Facility

The relationships between the user, group agent and directory are shown below.

7.2.3 Loop control

The group agent is responsible for ensuring that loops do not occur during the point to point expansion of nested distribution lists. The protocol supports three methods for such loop control: *Remembering Message IDs, Inbound History Trace* and *Outbound History Trace.*

It was decided to employ all three methods of loop control in the implementations. The Message-ID solution requires that a group agent store the IDs of all messages that it has seen so that it can recognise any which are looping. The History Trace solutions require that the names of all group agents which have expanded a message be included in its header information and the protocol ensures that any group agent mentioned does not expand a message again. This information can be checked when a group agent receives a message (inbound) and when it resubmits it to the MHS (outbound).

The group agent has to use the existing X.400 header fields for storing trace and other information. This is achieved by encapsulating the message in a new header when it is expanded at its first group agent. Subsequent group agents in the expansion modify the fields in this new outer header. The table below summarises the way in which X.400 fields are used on the initial and subsequent expansions.

Defined Terms	
Incoming P2 Message	I.P2
Outgoing P2 Message	O.P2
Group Agent	Ga
Incoming P1 Message	I.P1
Outgoing P1 Message	O.P1

Header Mappings	
Field	Setting
P2 level for initial expansion	
O.P2.AuthorizingUsers	AMIGO List Expander VI
O.P2.Originator	I.P2.Originator
O.P2.Message-Id	New Message-Id
O.P2.AutoForward	False
O.P2.PrimaryRecipients	Ga.Name
O.P2.ReplyToUsers	I.P2.ReplyToUsers
O.P2.CopyRecipients	default if not group-agents contained in Ga.ReceivingMembers
P2 level for subsequent expansions	
O.P2.AuthorizingUsers	list expander
O.P2.Originator	I.P2.Originator
O.P2.Message-Id	New Message-Id
O.P2.AutoForward	False
O.P2.PrimaryRecipients	I.P2.PrimaryRecipients Ga.Name
O.P2.ReplyToUsers	I.P2.ReplyToUsers
O.P2.CopyRecipients	I.P2.CopyRecipients + group-agents contained in Ga.ReceivingMembers
P1 level	
O.P1.Originator	Ga.Auditor
O.P1.Recipient	Ga.Recipients
O.P1.DiscloseRecipients	False

This concludes the overview of the AMIGO distribution list protocol. The following sections describe how this and other protocols were implemented by different AMIGO partners. These descriptions concentrate on the construction of group agents and their connection to X.400 MHS as well as the implementation of supporting directory systems. The implementation of a human usable directory access protocol using MHS as a transport for directory queries is also described.

7.3 Nottingham Piloting

This section describes the implementation of the AMIGO distribution list facility at the University of Nottingham. The implementation uses the existing *PP* message system[JO87] and a local Directory implementation.

7.3.1 The Group Agent

The group agent is a P2 entity separate from the mail delivery system which is designed to work closely with both the mail system and the directory service. Its function is to take incoming messages as delivered by the message transport service; perform the expansion process and the resubmit the transformed message back into the message transport system for final delivery. As such, the Group Agent can be thought of as three functional parts: The MTS interface, the directory interface and the transforming process. These parts are further described in the next sections.

Architecture of PP

PP is designed as a generic mail transport system not only as an X.400 mail transport system. For some years to come, X.400 will need to interwork with different versions of X.400 and also with completely separate protocols, such as the DARPA RFC-822 protocol[Cro82].

The PP system consists of several discrete programs which combine together as shown in Figure 7.2. All input to the PP system enters through one of the input channels (I chan). These channels are responsible for converting different incoming protocols into a standard format for placing in the queue. The input channel therefore takes an incoming message in whatever protocol is used, and writes to a submission process (Message Submission) in a standardized protocol. The submission entity then checks that the addresses given are reachable, that the message is authorized to use this route, and that there is a conversion path between the incoming format and the outgoing format. These functions are the responsibility of the verification element. If all these criteria are met, the message is placed in the queue.

At some later stage, possibly after reformatting of the message has taken place, the message is extracted from the queue (Message Extraction) and a delivery attempt is made. This is achieved by an outbound channel program (O chan), which takes the standard PP protocol describing a message and transforms it into the appropriate protocol for that transport medium. If the message is successfully delivered by the outbound channel, then the message is removed from the queue. Otherwise the message is either left for another try, or returned if there is an irrecoverable error.

Occasionally, a protocol can perform both input and output functions in the same session. In this case, a special channel is needed which can switch

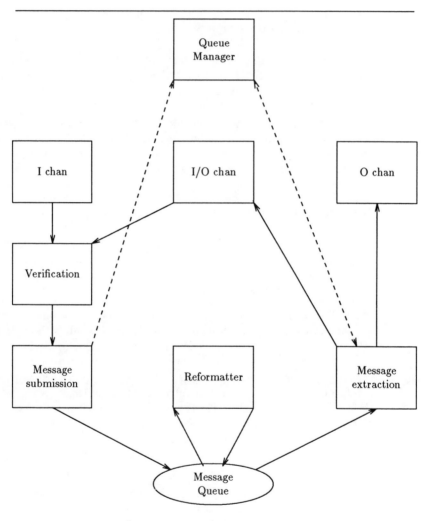

Figure 7.2: The PP Components

between submission and extraction (I/O chan). The AMIGO group agent is an example of an I/O channel as will be described below.

The Queue manager manages the interaction of the other programs, and oversees the running of the system. It schedules the startup of the various channels and invokes the reformatting processes where necessary. Each module communicates with the manager passing status information. In this way, this queue manager serves as an information and control centre, and does not have to do any work on messages themselves.

Routing of messages is handled by the submission entity using a database of information. The message is routed by the route with the least transformations.

The ISODE environment

The group agent, the directory and PP rely fairly heavily on a package of libraries collectively termed *The ISO Development Environment* (ISODE) [MTR87].

This is a package of routines that implement the transport, session and presentation layers of the OSI network model. It relies on existing network layers and at present has interfaces to four implementations of X.25, a TCP/IP network layer, and a hybrid TCP/IP-X25 bridge protocol. Several of these interfaces can be run simultaneously. This makes testing of the protocols over Local and Wide are networks very easy as the same interface is used for both.

At the higher layers, extensive use is made of a utility provided with the ISODE package termed *pepy*. This utility takes ASN.1 definitions and builds either a parser, a generator or a pretty-printer for the definitions. The ASN.1 definitions can be annotated with C statements and expressions at various points to allow values of C variables to be captured or encoded. In this way ASN.1 data structures can be manipulated within C programs and used in calls to ROS or RTS routines to pass data across connections.

The Group Agent and PP

The group agent can be viewed as an I/O channel in terms of Figure 7.2. PP invokes the group agent (channel) on receipt of messages for the distribution list, the group agent manipulates the message and then invokes the submission entity passing the message back to the MTS for delivery. The architecture of PP allows modules such as the group agent which accept messages, process them and resubmit them to be added to the mail environment with ease.

The MTS interface

The MTS delivery interface is relatively straight forward. The group agent I/O channel is invoked by the Queue Manager on receipt of a message destined for the group agent (in practice, it may batch up several requests and submit them together). The group agent will thus receive one or more X.400 messages to process.

As a prerequisite to expansion, the group agent must parse the P1 and P2 information to determine relevant submission details for later directory queries. This is achieved using the *pepy* ASN.1 parser generator which allows the P1 to be parsed and the objects of interest to be extracted.

At the same time as the information for the directory request is gathered, other extracted information is also stored. In this protocol, much of the outgoing message is constructed from elements of the incoming message. For instance, if the message has not been expanded by a previous group agent, the message must be encapsulated. This is easily done while parsing the original message.

Directory Interaction

After the relevant facts have been gathered from the incoming message, the expansion process can begin. It commences by the group agent connecting to the directory service using the *Remote Operations Service* and identifying itself. After this, the membership of the submitting entity is tested by a *Verify-Membership* operation. If the submitting entity proves to be a valid submitting member of the group then the receiving members are obtained from the directory by a *List-Membership* operation. Alternatively, if the submitting entity is not a valid member of the group submitting members the moderator of the group is read by a *Describe-List* operation. This information is stored for later use as the P1 receivers protocol element. The interface to the directory service is provided by a library of routines included in the group agent code.

After the destination of the message is obtained, other fields can be filled in. The *replytousers* field is another field that may need to be retrieved from the directory. When the optional information is gathered the final information required from the directory is the name of the group auditor. This is required for the P1 *originator* field, to ensure delivery reports are returned to the correct place.

The association with the directory can then be terminated however, if the group agent is processing more than one message it makes sense to keep the directory association established.

The Transformation

At this point, the group agent has the constituent parts required to reassemble the message. The message is assembled using pepy, this time generating an ASN.1 compiler for the P1 and P2 parts. As part of the ASN.1 compilation, the information gained from the directory is placed in the appropriate fields. As mentioned above, many of the components are simply copied from the incoming message.

When the message has been constructed, all that remains is to submit it to the MTS to be delivered to its recipients.

7.3.2 The Nottingham Directory Service

This section describes the use of the Nottingham Directory Service to support the group agent and deal with queries via the P2 based directory access protocol. The basic architecture and functionality of the directory are presented followed by more detailed descriptions of its use by the AMIGO distribution list facility.

Architecture

The Nottingham directory service is designed as a general directory service to support applications other than the AMIGO distribution list service and AMIGO P2 based directory access protocol (examples of other services are *MHS* or *File transfer*). Thus, the directory supports a general data model and set of operations with enough flexibility to support many applications and the AMIGO view of distribution lists must be mapped to this directory data model (a description of this mapping is given in the AMIGO specification and summarized below).

The directory functionality is distributed between two classes of entity: The *Directory Service Agent* (DSA) which is a server responsible for maintaining the database of directory information and implementing directory operations and the *Directory User Agent* (DUA) which is the application interface to the directory and maps between the application view and the conceptual directory model. Each application program accesses the directory via a DUA (the group agent and P2 based access mechanism are examples of different applications requiring different directory interfaces). The current Nottingham directory is non-distributed and so many DUAs access a single DSA which maintains the entire information base. (A distributed service is planned for the near future where responsibility for information will be shared between many DSAs) The DUA and DSA communicate as client and server respectively using the *Directory Access Protocol* (DAP). Implementation of the DAP uses the ISO specified *Remote Operations Service* and supporting ISO stack. The relationships between DUAs and the DSA is shown in Figure 7.3.

The Directory Data Model

The following paragraphs present an overview of the directory data model supported by the Nottingham directory service. Real world objects (e.g. distribution lists) are represented by entries in the directory. Entries contain sets of attributes where each attribute has a type and set of values representing a distinct property of the object. For example, a distribution list might be represented by a single entry containing attributes of type *submitting member, receiving member, moderator* etc. Entries are related hierarchically based on organizational relationships (e.g. a list entry as a child of an organization entry) to form a tree called the *Directory Information Tree* (DIT). The DIT specifies a hierarchically structured global name space with each object having a unique *distinguished name* consisting of an ordered sequence of attribute *Type = Value* pairs. AMIGO distribution lists are named by O/R-names which must be mapped to distinguished names. This is easily achieved because O/R-names are hierarchical and can be mapped to distinguished names with attributes types such as *C, ADMD, PRMD ... etc.* For example, an AMIGO distribution list might have the directory name represented by:

```
/C=gb/ADMD=btmhs/PRMD=des/O=nott/OU=cs/PN.S=AMIGO piloting/
```

The following sections describe implementation of the Nottingham directory service in greater detail. The first describes the implementation of the group agent DUA and the P2 access DUA the second describes the implementation of the DSA using the RTI Ingres relational database.

The DUA

The DUA is the application interface to the directory service and maps the application view of data and actions to the directory data model and operations. For example, it might map the AMIGO list operation *Describe-List* to the directory operation *Read-Entry*. It is also responsible for managing dialogue with the DSA via the Remote Operations Service. The DUA can be divided into two elements responsible for these functions as described by Figure 7.4.

The Application Part

The *Application Part* of the DUA presents the application interface and is therefore different for each application. The AMIGO distribution list facility requires two application parts, one interfacing the directory to the group agent and the other mapping P2 text encoded directory requests to directory operations. The interface maps application specific operations to the

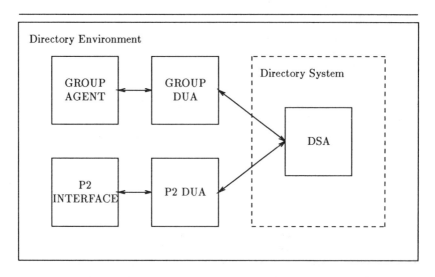

Figure 7.3: Relationship between AMIGO DUAs and DSA

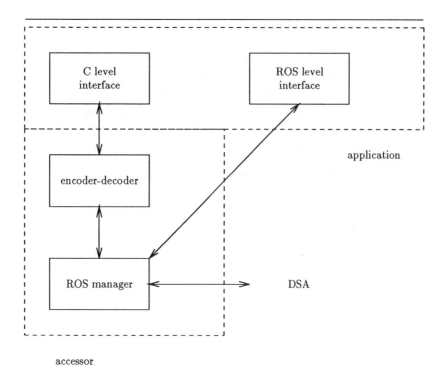

Figure 7.4: Internal structure of the DUA

Directory Access Protocol (DAP) which is specified in terms of ASN.1 data structures thus the application part builds arguments as ASN.1 structures, call operations directly via the Remote Operations Service (ROS) and interprets ASN.1 structures representing results. This interface to the ROS is represented bu the ROS level interface box in Figure 7.4.

A great deal of effort is required to build and parse ASN.1 structures and it would be wasteful to reproduce this work in each different application part thus the DUA includes a *encoder-decoder* which maps C representations of arguments and results to ASN.1 providing an additional C level interface to the directory as shown in the diagram.

The Accessor

The *Accessor* part of the DUA includes the encoder-decoder, routines to initiate and terminate an association with a DSA and a routine to call remote operations at the DSA. The Remote Operations Service and supporting OSI stack is provided by the *ISODE* package. The *encoder-decoder* is constructed using the *pepy* ASN.1 complier described in section 2.1.2. The accessor is common to all DUAs and consists of a library of C routines which are included with the code implementing the application part.

This paragraph outlines the application specific parts of the group agent and the P2 access DUAs. The application part of the group-dua is a library of routines included in the group agent code which map group agent operations to directory operations (e.g. *Verify-Membership* and *List-Membership* to *Read-Entry*). This library in turn includes the DUA accessor library.

The application part of the P2 DUA consists of a YACC [Joh78] compiler which parses the text of P2 encoded directory requests from its standard input and builds the arguments for directory operations. It prints the results of operations to its standard output. Thus the mail system can pipe the body part of a P2 directory request onto the P2 DUA and collect the output to be included in the return message.

The DSA

The DSA can also be divided into two elements: The *dispatcher* is responsible for implementing the DAP and is symmetrical in structure to the accessor in the DUA, the *executor* manipulates the information in the directory database and implements the directory operations. This structure is shown by Figure 7.5.

Directory information is stored in the *RTI Ingres* relational database management system (RDBMS). [Ing85]

A relational database was chosen for the prototype directory for the following reasons:

- It allowed the construction of the prototype within the available time.

- The presence of a ready made storage service allowed top down design of the directory to concentrate on the broader design issues.

- The relationship between directory services and RDBMS was worth exploring due to the number of RDBMS in operation today.

The executor provides a mapping between the directory data model (i.e. the DIT, entries and attributes) and the relational data model. This mapping is the connection between the elements of the executor as shown in Figure 7.5. The query language used with the Ingres RDBMS is *EQUEL* and the mapping is implemented by an Ingres programming interface called *embedded EQUEL/C* which allows EQUEL database queries to be embedded in and interact with C programs.

The executor is also responsible for transaction control within directory operations. Each operation is viewed as a single transaction which either completely succeeds or fails. For example, a modify operation may modify several attributes within an entry. The transaction control mechanism determines whether any individual attribute modification fails and if so rolls the database back to its starting state and generates a suitable error message.

The following operations are used to manipulate AMIGO distribution lists within the Nottingham directory:

```
Read_Entry(name, attributes)

Modify_Entry(name, new attributes)
```

Read Entry returns the values of requested attributes from a named entry. This operation supports the *DS-Verify-Membership* and *DS-Read-Entry* in the group agent and the *describe, expand* and *check* operations in the P2 based access protocol.

Modify Entry allows new attributes to be added to a named entry and the values of existing attributes to be altered or deleted. This operation supports the *add, delete* and *modify* operations in the P2 based access protocol.

The DSA implements an access control mechanism which determines which entities may perform which operations on directory information. This mechanism allows many different flavours of distribution list to be represented in the directory. For example, an open list may be created where anyone can add themselves as a new member or a secret list may be created where members are not allowed to read the attributes representing other members.

7.3.3   Future Work

The *PP* message system is currently under test at Nottingham supporting ISO/CCITT 1984 X.400. It is planned to extend PP to conform with
1988 X.400 and integrate it into other group communications environments
within the next year. The directory service is currently being expanded to
a distributed service with the naming tree distributed among many DSAs.

7.4   GMD piloting

This subsection describes the implementation of the AMIGO distribution list
protocol based on the EAN message handling service at GMD. It describes
the enhancement of the existing EAN distribution list facility to provide
group agent functionality.

The EAN message handling system produced by the University of British
Columbia, Canada, was used to realize distribution lists following the AMIGO
specifications. EAN was chosen for two main reasons:

> EAN already supports distribution lists where the expansion of
> a message sent to a list is performed by a special UA in the User
> Agent Layer. This corresponds to a P2 expansion agent similar
> to the proposed AMIGO *group agent* expansion method.

> GMD is serving as an EAN reference installation for several
> operating systems on behalf of the German Research Network
> (Deutsches Forschungsnetz (DFN)) and therefore a lot of knowl-
> edge about the internals of EAN is available within GMD. Work
> at GMD has also enhanced the EAN system to offer greater
> X.400 conformance and to support network management.

7.4.1   Distribution Lists in EAN

The original EAN (Version 2.0) includes the implementation of a simple
distribution list facility which is realized by a special user agent which we call
the *Group Agent* (GA) and which is responsible for distributing messages.
The GA is a P2 entity and sends all IP-messages which it receives to a set of
receiving members specified by a list contained in a file in the local system.

Unlike the AMIGO protocol the EAN protocol distributes the IP-message
without changes. Distribution is implemented by specifying the receiving
members of the list as recipients in the envelope of the message.

EAN distribution lists can be nested by specifying the names of GAs as
receiving members of lists however, no loop detection facilities are provided
in the EAN protocol and loops must be avoided by administrative means
(i.e. the list management facility should ensure that looping lists cannot be
created).

The EAN facility allows a single UA to be specified which receives all delivery reports produced as a result of distribution from GAs registered at the same MTA. In this case the name of this UA is used as originator for all distributed messages. In AMIGO terminology, this UA is a common auditor for all the GAs.

The distribution list facility is realized as a program named *dist* invoked whenever a message for the GA arrives. The *dist* program representing a GA interacts with an MTA as shown in Figure 7.6.

The GA and MTA interact in two sessions, firstly to deliver the inbound message to the GA, and secondly to resubmit the expanded message by the "common auditor" UA.

7.4.2   Extensions Implementing the AMIGO DL Protocol

The implementation of the AMIGO protocol consists mainly of the extension of the *expansion* step of the EAN protocol to include the stages shown in Figure 7.7.

The AMIGO implementation includes all three methods of loop control suggested in the protocol specification described earlier, namely *inbound history trace, outbound history trace* and *message ID*. These different methods of loop and duplicate control should prevent all cases (even unintentional) where the nested distribution lists contain loops.

In a first step the use of directory service is replaced by the use of the original EAN recipient file enlarged by the additional AMIGO attributes. These are interpreted as comments by the original EAN distribution program, to facilitate a smooth change to the new protocol.

7.4.3   First Experiences

The AMIGO protocol distribution program has now replaced the original EAN program for the distribution of messages to the members of the AMIGO MHS+ group. It has shown its usefulness when both versions were running simultaneously and there was an unexpected feedback situation between both group agents and a gateway. Whereas the traffic between the old GA and the gateway was growing geometrically, the feedbacks on the paths leading over the new agent were nipped in the bud.

7.4.4   Future Plans

After a period of test use of the AMIGO protocol distribution program we are willing to distribute the program to other interested EAN V.2 sites so that more users can profit by its advantages.

In parallel to this implementation, another project at GMD has implemented

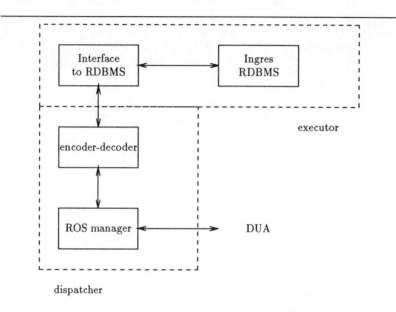

Figure 7.5: Internal structure of the DSA

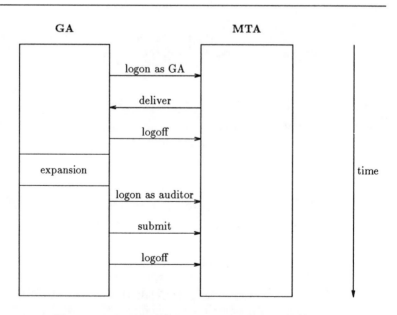

Figure 7.6: Interaction between the GA and MTA

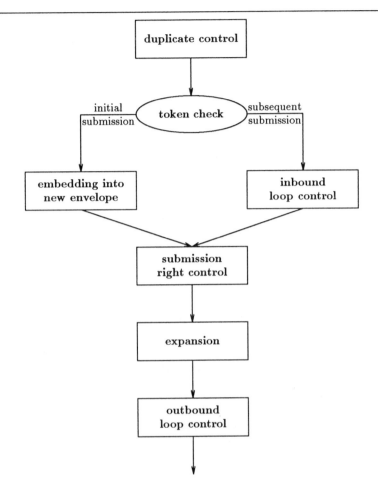

Figure 7.7: Functional entities for AMIGO Distribution List Handling

a Directory in the context of an MHS implementation by DANET (a German software house) on behalf of DFN.

We have planned to replace the current use of the distribution list file by the use of this directory conforming to the AMIGO pilot directory protocol.

Recent experiences and supplements to the protocol

The experiences with moderated distribution lists have lead to enhancements to the AMIGO protocol for a better support of the moderator role.

Moderator receipt of a message In most cases the moderator of a list is also a receiving member. If he receives a message from the list he needs an indication in which role (moderator or receiving member) he receives the message.

In the case of *moderator receipt* he has to decide on the resubmission of the message and possibly to resubmit it. If he gets the message as a *receiving member,* no action has to be taken.

To indicate the case of moderator receipt, the AMIGO protocol has been modified in our implementation. In the case where the submitter of the message is not a submitting member of the list, the ORdescriptor of the moderator is added to the P2 PrimaryRecipients (history trace) field before the GA sends the message to the moderator. The moderator can find out in which role he gets the message by checking this field.

Moderator submission   If the moderator (in this role) receives a message and decides to resubmit it, all trace information for loop control as well as the Message-ID are lost since the message gets a new P2 header.

To allow the resubmission of the unchanged original message the special case of *moderator submission expansion* has been introduced in addition to *initial* and *subsequent expansion.*

The moderator *forwards* the original message to the group agent (as the only body part) and indicates this fact by the FreeformName "moderator-submission" in the AuthorizingUsers field (of the outermost header).

The group agent checks this field in order to find out the expansion type. In the case of moderator submission he "unpacks" the forwarded message and then acts on the original message.

7.5   QZ Piloting

QZ has implemented the AMIGO piloting protocols for directory service access. This implementation was done for the COM computer conference system at QZ.

Since COM already had a built-in directory system for local COM users, no new directory system was implemented. Instead, an interface was developed to allow remote users to access the existing COM directory system using the AMIGO directory protocols.

7.5.1   Services Provided

Only a subset (plus a few extensions) of the full AMIGO protocols were implemented. The following functions were implemented:

DESCRIBE operation properties NAME, DESCRIPTION, CHARGING, MODERATOR.

DESCRIBE operation will not only find mailing lists, but also personal mailboxes.

ADD and DELETE operations, but only with flag REC and only for adding or removing yourself from a distribution list, and only for adding or removing members in open (public) conferences.

An additional command HELP, originally not defined by AMIGO, is available. It returns a user-oriented description of the directory services. If a parameter to HELP is given on the same line, for example HELP DTEADR, then for some parameter values, additional help is returned. The HELP command was later accepted by AMIGO as an extension of its protocol.

If the <list-name> operand of the DESCRIBE operation does not match one single directory name, a search is done for names of which <list-name> is an abbreviation. For example, the <list-name> "AMIGO" will find all names containing the word "AMIGO" as part of the name. This means that multiple values may be returned.

DESCRIBE will only find names of mailboxes and distribution lists residing in the MTA to which the command is sent. Even though the COM directory contains also non-local names, these will not be found.

An additional numerical parameter is allowed after "REC". Example: "ADD (AMIGO_group * REC 9)". If this parameter is given a non-zero value, then the new member of the distribution list will not only get forthcoming messages, but also the latest about 9 previous messages, if available in the COM data base, will be sent to the new member.

7.5.2 Experience

The directory service has been in production usage since September 1987. More than a thousand messages has been sent and received by the directory service agent during the first six months of operation. Users in ARPANET, BITNET, UUCP, MHS and other nets have been using the services.

7.6 Madrid Piloting

7.6.1 Introduction

The Multiuser Message Store is a system which provides storage services to many different users. Access is obtained by using an existing MHS (EAN) which extends the usage to all the users of this service.

This pilot implementation consists of two different parts: a Message Store Module and an Interface to the EAN system.

7.6.2   Message Store Module

From the organisational point of view, the MS is structured as messages and sets of messages which allow user defined organisation of information. An IP Message is considered as a particular case of a document and is represented by means of attributes.

Operation on messages are based on those attributes, which allow classification and easy retrieval (using filters). The implementation consists of a set of functions which handle a simple data base.

7.6.3   Interface to an EAN system

The Interface receives queries from the EAN users, performs operations on the MS and sends the answer back to the requestor. Both queries and responses are embedded in IP Messages. Operations to be carried out on the MS are codified in human readable form. In this way, the users needn't have any special agent or tool for getting access to the MS.

The Interface itself contains a syntactic analyser which is able to determine the correctness of operations syntaxes and to extract the actual information to be used as parameters of the real operations performed in the MS.

7.6.4   Operations structure

The structure of Store Operations embedded in IP Messages (as shown in Figure 7.8) is composed of:

- Query Header: some fields from this header are used to identify the initiator and to compose the reply messages containing responses.

- First Body Part. Contain operations and their parameters.

- Forwarded Message: Only exists in the *Store operation and will contain the full message to be stored.*

7.6.5   Multiuser characteristic of the MS

The MS allows access to multiple users who share information. To guarantee privacy and security of information contained in the MS, an Access Control mechanism has been added. This mechanism based on Access Control Lists can be applied to individual messages and to sets of messages.

7.6.6   Capturing Messages

To support simple group facilities (such as bulletin board or conferences) it is possible to combine distribution lists and the MS. In the pilot imple-

```
┌─────────────────────────────────────────────┐
│  Message inbox:7 - Sent                      │  Query
│  From:    CARLOS <carlos@euittm.iris>        │  header
│    To:    <MS@euittm.iris>                   │
│  Subject: Store a message in the MS          │
├─────────────────────────────────────────────┤
│                                              │  Operations
│    STORE_MESS <'MS'>;                        │  and
│                                              │  parameters
│  ┌────────────────────────────────────────┐ │
│  │ ====================                    │ │
│  │ Delivery-date: Wednesday, May 27, 1987  │ │  Header of the
│  │ From: MIGUEL <miguel@euittm.iris>       │ │  message to be
│  │ To:    <carlos@euittm.iris>             │ │  stored
│  │ Message_ID: inbox:6                     │ │
│  │ Subject: Sets within the Store          │ │
│  ├────────────────────────────────────────┤ │
│  │ In the MS a Document is defined like a  │ │
│  │ collection of attributes. The message   │ │
│  │ handling in MS Data Base is based on    │ │
│  │ these attributes.                       │ │
│  │                                         │ │
│  │     The attributes represent the        │ │  Body of the
│  │ greatest part of the knowledge that the │ │  message to be
│  │ MS has about the message. That knowl-   │ │  stored
│  │ edge is completed with information about│ │
│  │ operations performed on it. The search  │ │
│  │ and retrieval operations are based on   │ │
│  │ attribute value(s).                     │ │
│  │                                         │ │
│  │     An attribute ...                    │ │
│  └────────────────────────────────────────┘ │
└─────────────────────────────────────────────┘
```

Figure 7.8: Example of message to be stored

mentation, the name of an entity is included in the group distribution list and this carries out operations on the MS to store messages sent through the Distribution List. In this way, all messages distributed by the AMIGO Group Lists have been stored in the MS since the summer of 1987.

7.6.7 Conclusions

The experiences resulting from the pilot implementation give rise to the following conclusions:

- the structuring of messages into sets and compound objects facilitates the storage and retrieval of large volumes of information.

- the absence of interaction between the users and the MS produces a great delay in the natural sequence of search and retrieve operations for selecting required information.

- In the pilot implementation, operations are manually composed by the users. This is the main source of errors and delays. The complexity of operations represents a heavy burden for the users. In particular, the raw structure of Access Control Lists is very unfriendly.

Future piloting would have to be carried out by implementing the final AMIGO Data Model and operations. This piloting must take account of the above conclusions and try to correct those problems detected in the current implementation.

Chapter 8

Support for Advanced Group Communication

Authors: Bernd Wagner
           Jacob Palme

Previous chapters have described the extension of current communication services to support added functionality for group communication. In effect, the Amigo MHS+ approach is to place a new "group communication layer" on top of existing layered services.

Other projects in this area have adopted less pragmatic approaches and have developed highly abstract models for describing group communication in service independent ways. The Amigo MHS+ model can be viewed as an attempt to bridge the gap between these advanced models and existing services. In terms of a layered approach, it resides between them. Consequently, this chapter explores the relationship between these models for "advanced group communication" and the Amigo MHS+ models specified by previous chapters.

8.1   Knowledge Based Information Handling

Need for knowledge based information handling support

The simplicity of providing, replicating and distributing informations by electronic communication media results in an abundance of available information. This leads to the problem of information overload of the users. A detailed study of this problem and of its handling in existing systems has been performed by Hiltz and Turoff [HT85].

There is a clear need for tools which help to deal with this overflow of information, e.g. to

- react to some kinds of information objects automatically or semi-automatically,

- extract the relevant information,

- concentrate on the relevant parts of the information objects,

- minimize unwanted redundancy of information,

- allow structured information handling (e.g. handling of information in context).

The task of reducing the amount of available information is in some respects more intricate than that of producing it: knowledge is required about what is "relevant information", "redundant information", " information context", "action to be taken".

Knowledge based information handling is necessary both for individual purposes and for group communication. However, multiplication of the amount of information is performed mostly in the group communication context (e.g. by distribution lists, conferences). This subchapter points out some of the issues related to this problem area and some of the existing approaches to their solution. However, there is no integrated approach that addresses all the problem areas.

Knowledge based methods could be applied at all points of group communication activity:

- archiving,

- retrieval,

- receipt,

- distribution

of information objects.

Examples for concrete knowledge based information handling actions are

- taking actions related to information objects,

- prioritizing and filtering,

- categorizing and structuring,

- collecting and condensing.

Approaches to message management systems supporting some of these actions are described in Tsichritzis et al. [TRG+82], Malone et al. [MGT+87, MGL+87] (Information Lens) and Chang and Leung [CL87] (KMMS).

They employ structuring or "semi-"structuring of messages through the use of header fields. (In terms of the AMIGO data model, these are attributes of Information Objects.) Rules for the actions to take and the filtering criteria are defined in terms of these fields (attributes).

Taking Actions

The systems mentioned above support the definition of actions to be taken on receipt of a message.

Example ([MGL+87]):

IF From: Silk, Siegel
THEN *Set Characteristic*: VIP
    IF Message Type: Action Request
        Characteristics: VIP
THEN *Moveto*: Urgent

The Activity Model of AMIGO Advanced [DPB88] provides a powerful formalism to describe more complex actions to be taken in a group communication activity.

Filtering, Prioritizing

A detailed description of different categories of filtering criteria can be found in [MGL+87]:

- cognitive filtering (depending typically on complex combinations of keywords),

- social filtering (depending typically on the personal and organizational interrelationships of the communicating persons),

- economic filtering (depending of communication costs).

In [CL87] Chang and Leung give explicit filtering algorithms.

Their first case is filtering dependent of traffic parameters ("economic filtering" in the terms of above).

Their second algorithm is a combination of "economic" and "cognitive" filtering: The "relevance" of a message is defined in terms of user and message profiles (related to keywords). Prioritization is performed using both traffic parameters and relevance.

Categorizing, Structuring

The effectiveness of the use of the attributes of an information object in the "action rules" or "filtering/prioritizing criteria" depends of the quality of the actual attribute structuring of the object.

In order to extract additional structures from the object, text analysis is required.

Examples for approaches to structuring and classification of texts are in [EK88] and [HR88].

Collecting, Condensing

In conference systems there are often multiple replies to one contribution which are generated independently of one another and contain duplicate information. Condensation of this information is performed manually by a moderator (if it is done at all). Tools are required to support such condensing processes. In cases where the objects to be collected are very formalized, the collecting and condensing process can be automated (e.g. in the case of voting).

8.2 Group Activity Communication Structures Support

This section of the chapter discusses three examples of communication *structures*: voting, joint editing and meeting scheduling. The examples are intended to indicate what kinds of issues need to be addressed to fully support typical communication structures, they do not contain fully worked out solutions.

8.2.1 Voting

Voting requires a voting agent to control the voting procedure. The voting agent might perform one or more of the following actions:

- Sending out the vote query to the voters.

- Collecting the votes.

- Counting the votes, and sending the result.

This requires protocols:

1. Between the person requesting the vote and the vote agent. This protocol is used to specify who is allowed to vote, what vote format

is permitted, how the votes are to be counted, if the individual votes are to be secret (balloting) or not, at what time the votes are to be counted, where the results are to be published etc.

2. Between the vote agent and the voters, specifying the query and the permitted format of the replies. For example, the permitted format may be one of the alternatives "Yes", "No" or "Abstain", or a natural number between 1 and 5. In some voting procedures, the voter may be allowed to add a textual explanation to the vote.

3. Between the vote agent and the voters, sending out reminders to those who have not replied to the vote, shortly before the end of the voting period.

4. Between the voters and the vote agent, sending in the vote.

5. For sending out the result to the voters.

These protocols could be embedded in messages using the P22 protocol of X.400/88, since that protocol allows extended P22 heading fields in which the information can be specified. For example, there could be a P22 heading field called "Choices" which lists the permitted reply to the vote query etc.

The Group Communication System could be used to distribute a vote query to members of a defined group. The vote agent could then check incoming votes against the Directory System to check if the voter was a member of the group which is allowed to vote.

The voting protocol should permit votes with more than one question to be answered (similar to questionnaires).

One could create a new instance of a vote agent for the handling of each vote question, or one could let the Group Communication Agent provide voting services instead of having separate vote agents.

8.2.2 Joint editing

Joint editing is a procedure for producing a document by a group of authors communicating via the Group Communication System.

Joint editing requires a master document. Whether this requires that the master document is stored on one single host is for further study.

In order to make it easy for different people to work on different parts of the master document, it could be structured into a detailed hierarchical structure.

Each node in the hierarchical structure could be handled in much the same way as a message. In particular:

- It should have a globally unique ID, to allow unambiguous references to it, just like messages.

- It should be able to refer to other document nodes and/or messages just as messages can refer to each other with InReplyTo, References, and Obsoletes heading fields. Of special value for joint editing might be fields like:

  - Part of, to connect the nodes in a hierarchical structure representing the whole document being edited.
  - Obsoletes, to define the history of the document development.

- Ordinary messages should be able to refer to parts of the document. This can be used both to comment on, and discuss, a particular part of the document, and to transport suggested, but not yet approved changes to the document. A new heading field "SuggestedRevisionOf" might be useful here.

These additions can easily be added to the P22 protocols of X.400/88, using the extension facility built into that protocol for adding new heading fields. No real time connections to a single host holding the master copy are necessary.

Conferences/mailing lists could be set up to distribute changes to, and discussion on, the document. Separate lists may be necessary for different parts of the document.

An access control procedure is necessary to control who is allowed to make changes to the master copy of the document being produced. This might be different for different parts of the document.

A data base for information about a document in development might be useful. Such a data base might store:

- The master document.

- The updated history of the master document, through which older or alternate versions can be found.

- Development status of each part of the final document.

- Who is currently responsible for what part of the final document.

- Who is currently working on what part of the final document.

- Existing alternate versions of, and revision proposals for, each part of the document.

8.2.3   Meeting Scheduling

Theoretically, meeting scheduling could be accomplished by having a data base, containing the personal calendars of each participant. A computer could then, automatically, find a suitable meeting date and schedule the meeting in the calendars of all the participants.

In practice, this solution will not be easily realizable in a distributed environment. An alternative would be to develop a protocol for interaction between the participants trying to find a suitable date. Logically, this could be seen as an interaction between the data bases kept by each participant holding his/her calendar. However, all participants may not choose to keep their calendars in computer-processable format, and some participants may have conditions on the suitable date for the meeting which are not easily formatted in computer-readable terms.

Thus, the end users in this protocol will be humans, and the meeting scheduling must include the possibility of sending human readable messages between the participants. These human readable messages can be augmented by special, extended P22 heading fields to support exchange of the special data needed to schedule the data.

Note that the unwritten rules controlling meeting scheduling in a face-to-face meeting will not work when the scheduling process is handled via asynchronous communication (message-exchange). The reason why people often find meeting scheduling via message systems difficult is probably that they try to apply unsuitable protocols.

In particular, the protocol for meeting scheduling must:

- Include a minimum number of interactions. This means that more information must be transmitted in each interaction. Not discuss the suitability of one single data at a time, but consider in parallel several alternative possible dates.

- Acknowledge the fact that even though a participant has indicated that a certain date is free for him/her, this may not be true later on during the scheduling process.

The scheduling process might be aided by a scheduling agent which combines the information from the participants of the possible dates, and sends out summaries of which dates are still open etc. But the protocol should not be designed so that all communication must go via this scheduling agent. Messages directly between the participants of the group must also be allowed.

Appendix A

Existing Services

Authors   Manuel Medina
              Steve Benford

This appendix outlines four OSI application layer services that can be used to support basic group communication activities. Only a brief introduction to the four services is given, fuller descriptions can be found in the set of standard recommendations issued by the appropriate standard bodies. The following services are examined:

- The ISO MOTIS/CCITT X.400 interpersonal message service (MHS)

- The ISO/CCITT X.500 directory service (DS)

- The ISO File Transfer Access and Management Service (FTAM)

- The Document Filing and Retrieval Service (DFR)

Not all of these services are yet fully standardized. Even where standard recommendations exist, improvements are actively being contemplated. Furthermore, there are other on-going areas where potentially relevant models or services are being developed that are not covered by the MHS+ report due to their being at too early a stage of development. These include:

- DOAM - The ISO Distributed Office Architecture and Management model. This proposes a general framework for allowing disparate office services to interwork.

- DTAM - The CCITT Document Transfer and Manipulation standard developed by Study Group VII which is designed to allow the transfer of videotex and Office Document Architecture (ODA) documents.

- RDT - An ISO SC18 proposal for Referenced Data Transfer facilities that allow a way of transporting references to a document without having to transport the whole document where parties have equally good access to a common document filing and retrieval service.

The description of the four services is preceded by an introduction to Abstract Service Definition Conventions. These conventions are used in describing the MOTIS X.400 interpersonal message and several other services. They are also employed in the this report to describe parts of the AMIGO MHS+ group communication architecture.

A.1 Abstract Service Definition Conventions

This section outlines a set of conventions that may be used to describe Distributed Information Processing systems [ISO87a]. The intention is to simplify the design of a complex distributed information processing task by first allowing it to be described in an abstract way and thus ensuring that the specification is independent of any physical (i.e. real world) constraints. The following explanation may best be understood by reference to the illustrations in the following section describing the MOTIS interpersonal message handling service.

In these conventions, a Distributed Information Processing (DIP) task is described through two views: a macroscopic and microscopic view. The former description is called the *abstract model* and the latter the *abstract services*.

A.1.1 Abstract Models

A macroscopic description of a DIP task is called a model and it is based upon the concepts of abstract objects, ports, services and refinements.

An abstract object is a functional entity (component) of the DIP model which interacts with other functional entities to achieve the task. It is defined through a specification known as an *object macro*, which consists of a list of ports, and the roles the object plays in each of the ports.

A port is defined as a point at which an abstract object interacts with another abstract object. It determines the kinds of interactions, i.e. different physical links, channels or queues through which the object can interact. Ports may be symmetric or asymmetric, and in the latter case, one or two sided (single or bidirectional). The roles in an asymmetric port are identified generically as consumer and supplier. In a bidirectional port, both sides can perform both functions. Objects interact with each other through bound ports, and this binding operation can be performed only between two *matching ports*, i.e. two ports of objects with complementary roles.

The set of capabilities offered by one object to another, through one or more ports, is known as the abstract service. The former object is the service *provider* and the latter object the service *user*.

In some cases, the external description of an object is not clear enough to explain its functionality. In this case it might be more convenient to think of the object as being composite (i.e. composed of other objects). The functional decomposition of an object into several lesser objects is called the *refinement*. The technique of refinement can be applied recursively until one reaches component objects that can be best considered atomic.

Refinement Specification Structure

The specification of a refinement structure has two parts, like a program in most programming languages. The first part is a list of definitions of the objects, services and ports used in the description of the refinement of the module (e.g. see Figure A.7). In that part, there is also a list of the identifiers *exported* to other modules. This means that some of the definitions needed for the refinement are tagged as *imported*, i.e. the complete specification of that object or service is given in another module.

The second part is the refinement specification itself. As already mentioned, the refinement is expressed as a list of components. Each component is an object and the list of its ports, with the indication of either the object to which the port is bound, or that the port is *visible* (i.e., bound to the external world view). The visible ports are those through which the specified object interacts with other objects, when it is used in other refinements.

The main advantage of this notation is that it allows one to describe the rules to bound the objects, rather than a particular configuration. This is possible thanks to the capability of specifying an object as *recurring* - meaning that the object can appear inside the object refined as many times as required in a particular instantiation. (See Figure A.1).

On the other hand, there is no easy way to specify that an object, or one of its ports, is optional and therefore can be omitted in some of the instantiations of the object being refined. One possible interpretation of this lack of capabilities in the description language is that, in most cases, the absence of one object or port in an object refinement should be associated with a different functional profile of that object. This means that we will have families of objects, some of them supersets of others (i.e. performing a superset of the functions of others) if they include components and/or ports not available in the others.

A.1.2 Abstract Service Definition

A microscopic description of a DIP task is a specification of the abstract service that defines how the task is initiated, controlled, and terminated. It is based on the concepts of *abstract bind operations*, *unbind operations operations*, and *errors*

The service offered by an object is defined as the set of operations that can be performed by that object at another's request, through the set of ports that populate its surface. The tasks performed as a result of the invocation of an operation from another object are known as a *procedure*. Procedures may have a single argument of a prescribed type, and may produce as outcome a single information object: the result, if fully completed (i.e. succeeded); or certain information, if prematurely terminated (i.e. failed).

The services provided by any object always include the basic operations *bind*, *unbind*, and *error*. The difference between two services resides in the definition of the data type of the argument, and the possible error information and result.

In addition to these basic operations, we may have zero or more operations specific to the service. These are of the *operation* abstract type, and some of them may be common to several ports of the service. They may have different identifiers, but the restrictions about the uniqueness of argument and result information objects are also applicable to them.

The information objects involved in the operations are defined using the ASN.1 notation[ISO87b] that allows one to express choices among several data structures, depending on some of the fields. In practice, this allows one to have several operations included in only one definition.

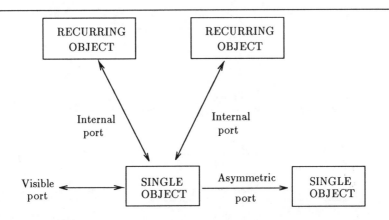

Figure A.1: Very simple environment showing the most important elements used to describe a model.

A.2   Message Oriented Text Interchange System (MOTIS)

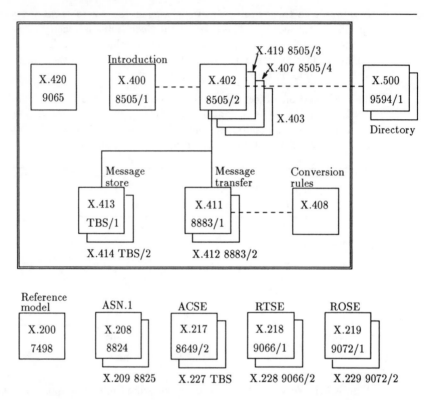

Figure A.2: Message handling and foundational specifications

In 1984 CCITT published a set of interpersonal message service recommendations known as the MHS X.400 series. Subsequently, CCITT and ISO started to unify their efforts in this area and in 1988 they produced a common MOTIS/MHS set of recommendations. There are substantial differences between the 1984 and 1988 versions. At the time of writing there are a few 1984 systems being offered by vendors but no fully 1988 compliant systems. The outline description in this appendix relates to the 1988 MOTIS/MHS version.

There are a large set of recommendation documents that serve to fully define the MOTIS service. As a reference guide for the reader, the structure of these recommendations is provided in Figure A.2.

A.2.1   MOTIS Architecture

The MOTIS architecture allows one user to transfer a message to another user in a variety of ways. The transfer may be achieved directly, or it may be use intervening facilities. The facilities include the use of a storage unit for message deferral, and the use of a distribution list expander which conveniently allows one-to-many message transfer.

The architecture is shown in a set of three figures that follow a top-down approach. These cover the user view, message system view, and the message transfer system view. Figure A.3 shows the outermost view of the system, i.e. the view that the user has.

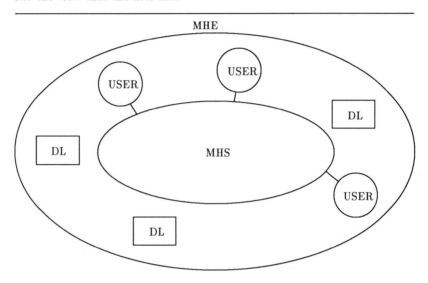

Figure A.3: Message Handling Environment

In the figure, DL stands for Distribution List, and represents a group of users (which may contain other DL as its members). The MHS represents the distributed system that allows the users to send, receive, and store messages, and also manage the messaging facilities.

The *Ports* through which the user accesses the system are shown in Figure A.4.

origination **PORT**
   CONSUMER INVOKES{
      OriginateProbe,
      OriginateIPM,
      OriginateRN,
      CancelIPM}
   ::=id−pt−origination

reception **PORT**
   SUPLIER INVOKES {   10
      ReceiveReport,
      ReceiveIPM,
      ReceiveRN,
      ReceiveNRN}
   ::=id−pt−reception

management **PORT**
   CONSUMER INVOKES {
      ChangeAutoDiscard,
      ChangeAutAcknowledgment,   20
      ChangeAutoForwarding}
   ::=id−pt−management

Figure A.4: User interface through ports

Inside the MHS object, there are the components shown in the Figure A.5. The User Agents (UA) allow the users to access the Message Transfer System (MTS), either directly or through the Message Store (MS) that allows the more or less transitory storage of messages before being read by the user. The UA uses the MTS through three *Port* definitions which appear in Figure A.7.

The MTS and the UA use the MS through the *Submission* and *Administration* Ports. In addition the MTS also uses the *Delivery* Port, and the UA uses the *Retrieval* Port (see Figure A.6).

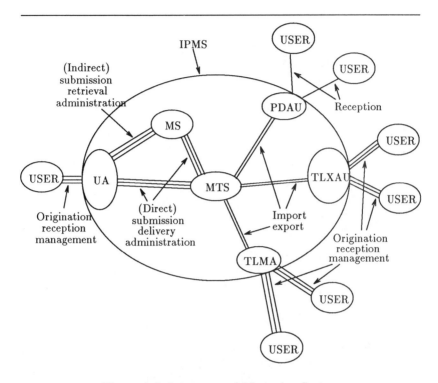

Figure A.5: Interpersonal Messaging System

Retrieval **PORT**
    BIND AuthenticateUser
    SUPPLIER PERFORMS {
        Summarize, List, Fetch,
        Delete, Register−MS}
    CONSUMER PERFORMS {Alert}

Figure A.6: Retrieval Port

−−*Message Handling Environment refinement*
mhe−refinement **REFINE** mhe **AS**
    user **RECURRING**
        origination [C] **PAIRED WITH** {MHS}
        reception   [C] **PAIRED WITH** {MHS}
        management [C] **PAIRED WITH** {MHS}
    DL **RECURRING** MHS
    ::= id−ref−primary

SubmissionPort **PORT**   10
    SUPPLIER PERFORMS {
        MessageSubmission, ProbeSubmission, CancelDeferredDelivey}
    CONSUMER PERFORMS {SubmissionControl}
    ::= objIds−ports−submission

DeliveryPort **PORT**
    SUPPLIER PERFORMS {DeliveryControl}
    CONSUMER PERFORMS {MessageDelivery, ReportDelivery}
    ::= objIds−ports−delivery

                                                    20

AdministrationPort **PORT**
    SUPPLIER PERFORMS {ChangeCredentials, Register}
    CONSUMER PERFORMS {ChangeCredentials}
    ::= objIds−ports−administration

ipms−refinement **REFINE** ipms **AS**
    mts submission   [S] **PAIRED WITH** {ipms−ua, ipms−ms}
       delivery   [S] **PAIRED WITH** {ipms−ua, ipms−ms}
       administration [S] **PAIRED WITH** {ipms−ua, ipms−ms}
       import   [S] **PAIRED WITH** {tlma, tlxau, pdau}   30
       export   [S] **PAIRED WITH** {tlma, tlxau, pdau}
    ipms−ua **RECURRING**
        origination [S] **VISIBLE**
        reception   [S] **VISIBLE**
        management   [S] **VISIBLE**
    ipms−ms **RECURRING**
        submission   [S] **PAIRED WITH** {ipms−ua}
        retrieval   [S] **PAIRED WITH** {ipms−ua}
        administration [S] **PAIRED WITH** {ipms−ua}
    tlma **RECURRING**   40
        origination [S] **VISIBLE**
        reception   [S] **VISIBLE**
        management   [S] **VISIBLE**
    tlxau **RECURRING**
        origination [S] **VISIBLE**
        reception   [S] **VISIBLE**
        management   [S] **VISIBLE**
    pdau **RECURRING**
        reception   [S] **VISIBLE**
    ::= id−ref−secondary   50

Figure A.7: Interpersonal Messaging System

To reach the indirect users of the system, i.e. those without frequent access
to a terminal, or who are not subscribers of the MHS, we need to use another
system (e.g. ordinary mail, facsimile). To access these services various kinds
of Access Unit (AU) are defined: Physical Delivery (PD), Telex (TLX), and
Telematic (TLM) (see Figure A.5). These AUs use the MTS object through
the *Import Export* Ports (to be provided from the AU specification).

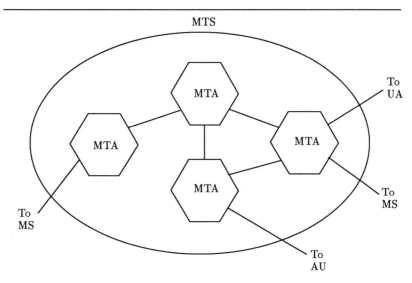

Figure A.8: Message Transfer System

Inside the MTS there is a set of Message Transfer Agents (MTAs), that
allow the transfer of messages from one user to another, in any location in
the world (Figure A.8). The links among MTAs represent the registration
of one with the other, i.e. the capability of direct transfer of messages
between them, without having to pass through another MTA, and this is
done through the *Transfer* Port. It is possible to send messages from any
MTA to any other, and intermediate MTAs depend only on the fact that
both MTAs know the X.25 and transport address of the other.

A.3   Directory Services

This section presents an overview of the first joint ISO/CCITT X.500 standard for Directory services, due to be ratified in 1988.

A.3.1   A brief history

The urgent need for a global Directory service has been recognised for some time. As the ISO's *Open Systems Interconnection* (OSI) model for computer networks is widely adopted in the near future along with various communications standards, such as X.400 for electronic mail, this need will become critical.

The general view is that the global Directory will be realised by an interconnection of Directory services provided by a combination of the national PTTs and private organisations. Both the *International Standards Organisation* (ISO) and the *International Telegraph and Telephone Consultative Committee* (CCITT) recognised the need for an international Directory standard to specify this interconnection and initiated standardisation work in the early 1980s. The early work of these committees was strongly influenced by ideas emerging from the *European Computer Manufacturers Association* (ECMA)[ECM87] and the *International Federation for Information Processing - working group 6.5* (IFIP 6.5)[IFI83] who, in turn, were influenced by existing systems such as the *Clearinghouse*[OD81] Consequently, early standards output resembled the Clearinghouse model and the output of the ISO and CCITT were similar. As a result, the ISO and CCITT merged their work in 1986[CCI86b] and have cooperated since that date to produce a joint standard. This standard is called *X.500.*

Due to the demanding timescale and the complexity of many directory issues, it appears that the 1988 version of X.500 will specify a fairly simple interconnection supporting the retrieval and limited modification of directory information. Issues concerning the management of information and the management of the service itself will be left open for the next study period.

A.3.2   The position of X.500 within OSI

The Directory service will play a vital role in the integration of communication services. The Directory service resides within the *application layer* (layer 7) of the OSI model and may utilise the supporting OSI stack for its internal operation and interaction with other services. More specifically, the distributed operation of the Directory may utilise the *Remote Operations Service*[CCI86c] and a number of presentation layer standards such as *ASN.1*[ISO86]. In addition, the Directory may need to interact with other future application services such as an *authentication service*[MNSS87, Bry88].

The interaction of the Directory with other services has been studied by a number of groups, including the ISO, who are producing a general model for *Distributed Office Applications*[ISO88] and the European Community COST 11-TER funded *AMIGO* project who have produced a distributed group communication model[BW88]. These models portray the Directory as a supporting service for a number of other services, interacting on a client-server basis.

A.3.3   Overview of X.500

The following pages summarise the major features of the X.500 standard in its present form. It should be noted that, although the standard appears to have achieved a stable state, it may be subject to future revision.

The terms *Directory* or *Directory service* refer to the X.500 Directory service for the remainder of this section.

Scope of X.500

The X.500 standard is described within an *Open Systems* environment allowing the interconnection of systems:

- From different manufacturers,

- Under different managements,

- Of different levels of complexity,

- Of different ages.

The Directory service is intended as an information service for OSI applications and is not a general purpose database system. The Directory has different operational requirements from such systems. For example, the rate of updates to information is expected to be far lower than the rate of retrievals.

For reasons of scale and management, the Directory is a distributed service provided by a number of physically separated application entities called Directory System Agents, each of which knows a part of the total directory information. However, from the user point of view, the logical results of directory operations are usually independent of their location.

The 1988 X.500 Directory service supports the following basic functionality:

- *Read* functionality. This includes the name to attributes mapping.

- *Search* functionality. This involves an attributes to set of names mapping.

- *Modify* functionality. This allows the limited update of information.

The X.500 standard is divided into several sections describing its *information model*, *protocols* and *distributed operations* as well several other issues. These are outlined in the following sections.

Information model

The directory *information model*[CCI88a] specifies the abstract structure of directory information. The Directory stores information about *communication entities* which are the humans, groups and application entities taking part in communication. Each communication entity is represented by an *entry* containing the information known about the entity. The set of all entries defines the total information stored by the Directory and is called the *Directory Information Base* (DIB).

Each entry consists of a set of *attributes* representing specific known facts about the entity. For example, an attribute might represent a mail address, a member of a distribution list or a textual description. Each attribute has an *attribute type*, indicating the type of information represented, and a *value*, containing the information. An entry may contain more than one attribute of a given type. Attribute types are globally unique, being represented by ASN.1 object identifiers.[1] A number of standard attribute types are defined in [CCI88b].

Entries are grouped into *object classes* specifying generic groupings based on the type of communication entity they represent (e.g. *organisational person* or *group of names*). Each entry contains a special attribute of type *object class* indicating to which object class the entry belongs. A number of standard object classes are defined in [CCI88c].

Table A.1 shows an example entry representing the organisational person *Fred Bloggs*.

Attribute type	Attribute value
common name	Fred Bloggs
surname	Bloggs
object class	organisational person
telephone number	+49 606 532776
user password	bananas
title	professor
...	

Table A.1: Example directory entry

The general structure of entries is shown as part of Figure A.9 below.

---

[1]Character strings are usually used in documentation for reasons of legibility.

Naming

The Directory service employs a hierarchical naming scheme for entries. Entries are arranged into a tree structure reflecting the organisational relationships between the communication entities they represent. This tree structure is called the *Directory Information Tree* (DIT) and is responsible for determining the Directory naming policy.

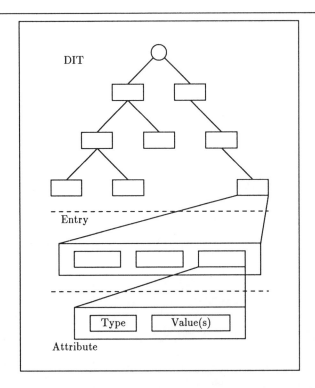

Figure A.9: General structure of the DIT and entries

Each vertex of the DIT is an entry, labeled with a *relative distinguished name* unambiguously identifying it among its siblings. The relative distinguished name (RDN) is composed of a subset of the entry's attributes called *distinguished attributes*. An entry's RDN is assigned by its naming authority, represented by its parent entry in the DIT. Thus, the responsibility for managing names is distributed throughout the DIT.

Each entry has a globally unique and unambiguous *distinguished name*, composed of the ordered sequence of RDNs encountered on the path from the root of the DIT to the entry. Distinguished names provide the basic handle on entries and their contents.

The relationship between the DIT and entries is shown in Figure A.9 and the relationship between the DIT, relative distinguished names and distinguished names is shown in Figure A.10.

DIT 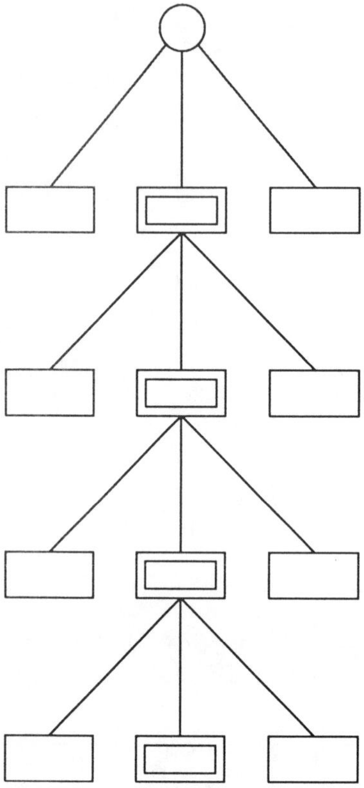	RDN	Distinguished Name
	C=GB	C=GB
	O=BT	C=GB,O=BT
	OU=Sales L=Ipswich	C=GB,O=BT, OU=Sales, L=Ipswich
	CN=Smith	C=GB,O=BT OU=Sales, L=Ipswich, CN=Smith

Figure A.10: The relationship between the DIT, RDNs and DNs

Relative distinguished names, and hence distinguished names, are generally chosen to be stable over long periods of time and also to be user friendly where possible. This requires the use of human guessable distinguished attributes.

A distinguished name need not be the only name for an entry. An alternative name, or *alias,* may be supported by the use of special pointer entries called *alias entries.* Alias entries do not contain any attributes other then their relative distinguished names and may only be leaf entries in the DIT.

A directory user names an entry by supplying an ordered set of purported

Abstract ports and operations	
port	operation
Read	Read
	Compare
	Abandon
Search	List
	Search
Modify	Modify Entry
	Add Entry
	Remove Entry
	Modify RDN

Table A.2: Directory Ports and Operations

attributes. These are mapped into the desired entry by the process of *name verification*, performing a distributed tree-walk through the DIT. Name verification provides the basic directory name to attribute mapping and indicates whether a name is valid or erroneous. A single name verification may dereference several aliases during its tree walk. Dereferencing replaces the attributes of the purported name matching an alias with those forming the name of the aliased entry.

The Abstract Service Definition

The directory *Abstract Service Definition*[CCI88d] specifies the user functionality of the Directory service in terms of a set of abstract ports and operations. The operations form the *Directory Access Protocol* (DAP) and provide the user with the functionality to retrieve, search and modify information. The ports and operations defined by the Abstract Service Definition are listed below and briefly described in Table A.2.

General read functionality is achieved via the *Read* port, supporting three operations.

- The *Read* operation returns the values of specified attributes from a single named entry.

- The *Compare* operation returns an indication of whether a named entry contains a specified attribute type/value pair.

- The *Abandon* operation allows the termination of those operations interrogating the Directory.

General browsing of directory information is achieved via the *Search* port, supporting two operations.

- The *List* operation returns the names of the children of a named entry.

- The *Search* operation supports the searching of DIT subtrees for entries matching specific patterns of attributes. The user names a subtree of the DIT, specifies some target attribute types and formulates
  an expression combining a number of attributes using the logical *and,
  or* and *not* operators. This expression is called a *filter*. The operation
  returns the values of the target attributes from those entries in the
  named subtree, matching the filter.

The limited modification of information is achieved via the *Modify* port.

- The *Modify Entry* operation adds, replaces or removes a number of
  attributes within a named entry.

- The *Add Entry* operation creates a new leaf entry within the DIT.

- The *Remove Entry* operation deletes a leaf entry from the DIT.

- The *Modify Relative Distinguished Name* operation alters the RDN of
  a named leaf entry.

It is important to note that the latter three operations only apply to entries
which will remain as DIT leaves. They do not provide a general facility for
building and manipulating the DIT.

Schemas

The structure of the Directory Information Base is governed by a set of
rules called *schemas.* These are integrity constraints ensuring that directory
information conforms to well defined formats. Schemas specify rules for the
following:

- The structure of names and hence the DIT.

- The contents of entries in terms of the attributes they contain.

- Permissible attribute types.

- The syntaxes of attribute values and rules for comparing them.

Each attribute in the Directory is governed by a rule assigning it a unique
Object Identifier and specifying its syntax. In addition, this rule states the
mechanism by which attributes of this type are compared with one another.

Each entry in the DIT belongs to an *object class,* governed by a schema.
This schema specifies *mandatory* and *optional* attributes for entries of this
class. Schemas may be nested, allowing more complex object classes to be
constructed from a few basic ones.

Naming rules govern which object classes may be children of which others
in the DIT and therefore determine possible name forms.

Standard attribute types and object classes are defined in [CCI88b] and [CCI88c] respectively and any directory operation attempting to violate these rules will fail. The relationship between schemas and the directory information framework is shown in Figure A.11.

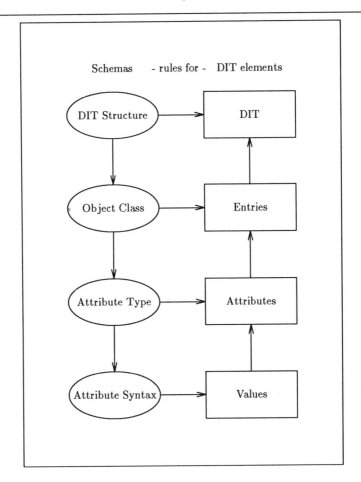

Figure A.11: The relationship between schemas and the DIT

Functional model and distributed operation

As mentioned above, the global Directory will be a distributed service. The procedures for its distributed operation are specified in [CCI88e].

The information constituting the Directory Information Base will be shared between a number of application entities called *Directory System Agents* (DSAs). These cooperate to perform operations, with each DSA knowing a fraction of the total directory information. DSAs can be viewed as a combination of local database functionality and remote interface to users and other DSAs. DSAs may cooperate in order to execute operations. Cooperation may take several forms and requires the navigation of operations through the distributed system. The set of all DSAs forms the *Directory system.*

A user accesses the Directory via an application entity called a *Directory User Agent* (DUA). DUAs manage associations with DSAs and present various interfaces to directory users (human or application). The provision of the Directory service by DUA and DSA functional entities is shown in Figure A.12.

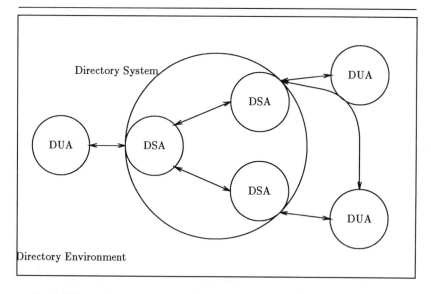

Figure A.12: The Directory provided by cooperating DUAs and DSAs

A user requests operations via their DUA. These operations are navigated through the Directory system until a set of DSAs is found which can perform them and return the results to the user.

DSAs may utilise several modes of interaction. These are *chaining, referrals (DUA or DSA based)* and *multicasting* as shown by Figure A.13.

Chaining occurs when DSAs recursively pass the operation to other DSAs which may be able to satisfy the request. The process of locating these *responsible* DSAs is called *navigation*. Once a satisfactory DSA is found, the operation is performed and the results are returned along the chain.

DUA referral involves a DUA contacting a sequence of DSAs to navigate the operation. The result of each contact is either the results and errors associated with the operation or the name and location of another DSA to be contacted (a *referral*).

DSA referral is similar to DUA referral except that a single DSA is responsible for handling referrals instead of a DUA. DSA interactions may also

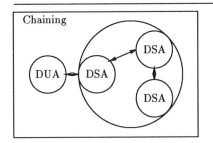 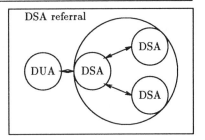

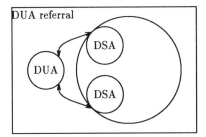

Figure A.13: Different modes of DSA interaction

involve *multicasting* where a query is decomposed into subqueries which are transmitted to several DSAs simultaneously. The results are then collected and merged before being returned to the user.

The standard allows DSAs to choose among these methods of interaction depending on conditions, their capabilities and policies.

A detailed description of the distributed operation of the Directory can be found in part 4 of the X.500 standard [CCI88e].

A.4   FTAM

A.4.1   Abstract Model

FTAM (File Transfer Access and Management) is an access and file management virtual system allowing any user to refer in a standard way to local and
remote files. It allows the user to transfer, access and manage the file contents, regardless of the remote hardware and software for file organization
and file management. Basically, FTAM is composed by two main elements:
the Virtual Filestore (VFS) and User File Service (UFS) modules.

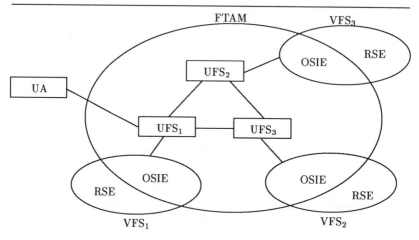

Figure A.14: FTAM abstract model

The VFS performs the application association between the Real System Environment (RSE), that contains the Real System particular characteristics,
and the Open System Interconnection Environment (OSIE), that allows the
interconnection of different systems. UFS, the User File Service, provides
the necessary set of Services, Facilities and the Protocols needed by a user
application to establish application associations with the remote systems in
order to access a particular Filestore.

A.4.2   Data Structures and Operations

The Virtual Filestore definition is actually a stored information description
method. In this description, a file is an entity with an identification name,
a set of general attributes, a set of particular attributes, and the data.

In a similar way, the definition of the User File Service provides a description
of the available type of services. In this description a Service is an activity
developed under an application association with a set of activity attributes
that identify that service.

A file contains one or more Data Units (DU). These DU's could be organized under different logic forms, such as sequential, hierarchic, net or relational. In FTAM, the DU's are organized in tree's form. Each node of the tree is able to contain some (or none) DU's. The unit where each operation is performed is the File Access Data Unit (FADU). The tree structure is named the Access Structure (see Figure A.15.

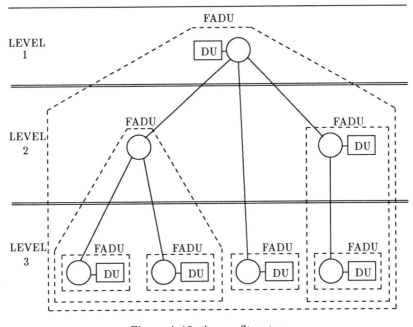

Figure A.15: Access Structure

Each DU contains one or more units of atomic data named Data Elements. Some special cases of data elements are single elements, tuples, arrays, matrix, etc.

The access structure is accessed in pre-order basis and each node is identified by the node list found when it is reached, starting from the root node. Each DU is associated to a node. Accordingly, each DU is identified with a DU list.

The DU attributes are classified in three types:

- scalar attributes: only can have one value at a time

- vector attributes: ordered list of some (or none) elements with a single value for each element

- set attributes : non-ordered set of some (or none) values

For each attribute the value type is also defined, i.e. a string of printable characters, a byte sequence, an integer value, an entry point address to a file service, a date, a time or an item from a named set of values.

As stated earlier, the attributes are grouped into three classes: general file attributes, particular file attributes and activity attributes. These attributes are divided in three main groups: Kernel, Store and Security.

- Kernel Group

  - File Attributes
    * filename
    * presentation context
    * access structure type
    * current filesize

  - Activity Attributes
    * requested access
    * location of initiator
    * current access structure type
    * current presentation context

- Store Group

  - File Attributes
    * account
    * date and time of creation
    * date and time of last modification
    * date and time of last read access
    * identity of creator
    * identity of last modifier
    * identity of last reader
    * file availability
    * permitted actions
    * future filesize

  - Activity Attributes
    * current account
    * current access context
    * concurrency control

- Security Group

  - File Attributes
    * access control
    * encryption name
    * legal qualification

  - Activity Attributes
    * identity of initiator
    * password

The file services and their protocols form an environment in which work is performed in a series of steps when the initiator runs activities. The dialogue must accordingly:

- allow the initiator and the responder to establish the respective identities.

- identify the file we want to operate.

- establish attributes describing the file structure we want to access.

- engage in file management or bulk data transfer.

These steps forms several parts of the application context. The period during which context information of some of those parts is valid is named the regime.

The interval during which a specific operation is performed, for example communication establishment, use or modification of the application context, is named the phase. For each phase, the valid message set is defined in terms of state transitions. In any interval, each entity is in one phase; one phase cannot be inside another phase.

The phases are divided in sub-classes, according the order in which they are executed during the file service utilization.

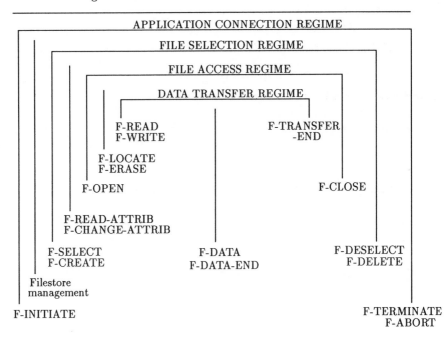

Figure A.16: Regimes

Each service composing the different regimes has the arguments shown in Figure A.16.

A.4.3   File identification and store navigation problems

The complete filename in FTAM (FADU) is composed of a filestore name and a filename. The filestore name identifies the filestore that contains the file (this name is optional for the local filestore), the filename identifies unambiguously the file in the filestore.

The filename is composed of a string vector (varying from zero to eight characters in each element) that determines the path to the file in the access structure and a vector of attributes that defines the particular file characteristics.

One problem is that in order to identify a file it is necessary to know the path to the file in the access structure. Other problems may be:

- a file can be selected only if it was not selected previously and allocated by another user.

- preorder must always be used to include information in the access structure.

- when a structure node is selected, we have access to all its sub-tree, so when the node is removed the subtree is also deleted.

In addition to the precedent problems related to the access structure, the system services and protocols do not allow the selection of more than a single file by application association. If we need to work simultaneously with more than one file, we need more than one application association.

A.5 Document Filing and Retrieval (DFR)

A.5.1 Overview of the System

The Abstract Model of the DFR is defined in terms of the Document Filing and Retrieval Environment (DFRE). The DFRE comprises primarily two types of functional objects: the DFR System and Users. The DFRS can be decomposed into "secondary" functional objects: DFR Servers and Document Stores. Figure A.17 shows the relation between the DFRS and its users.

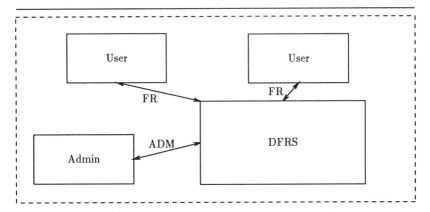

Figure A.17: Relationship between DFRS and its users

A user interacts with the DFRS by means of File, Retrieval and Administration Ports. The collection of capabilities provided by these Ports defines the Abstract Service of the DFRS.

A user acting as Administrator would use the Administration Port (the functionalities of which are left for further study).

A user get access to the DFRS by means of the "Abstract Bind Operation" and the DFRS uses the parameters of that operation for authenticating the user.

A.5.2 Information Model

The information Model is defined in terms of the structure of a Document Store. A Document Store is a collection of documents which can be arranged in a hierarchical structure as in Figure A.18. The Document Store is a named element which can be selected for operation by users.

The "atomic element" to be handled within the Document Store is the "Document", although documents can be grouped in "Groups". Furthermore, a document can belong to several groups by means of objects called "References". References can be internal or external.

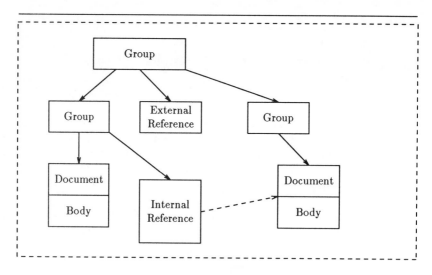

Figure A.18: Document Store

All the objects are subject to Access Control, but access control mechanisms are not fully defined yet.

A.5.3 Documents

A document consist of "Attributes" and "Content". The attributes are associated with the content. The content is the actual information to be stored in the DS (Document Store) and it is not currently interpreted by the DFRS.

Attributes are defined to qualify and describe the characteristics of the documents and its intended use.

A.5.4 Groups

A group is a collection of Documents (and/or References) to be handled as a whole and it is a named entity.

A group has a set of attributes associated with it and a set of members which can be: Documents, References and other Groups.

The creation of a Group can be performed by identifying the objects explicitly or by specifying a Criteria for membership based on an objects attributes.

The Criteria can be permanent (so it is used to govern the inclusion of new objects within the group) or can be only applied in the Group creation and no further effects will be produced.

A Group can be permanent or temporary (in the latter case it is only valid during the user's association with the Document Store).

A.5.5 References

A reference is primarily defined to allow a document to participate in more than one Group without requiring several distinct copies.

A Reference consist of a "Pointer" (specifying the location of the document) and a set of attributes associated with the Pointer.

Two type of References have been defined:

- Internal (referring to a Document within the same Document Store)

- External (referring to a Document in another DS)

The possibility of handling References to a Group has been left for further study.

The Reference Attributes are generally used to provide additional information about the referred document.

A.5.6 Naming of objects

All objects within the DFRS have an UPI (Unique Permanent Identifier). This UPI is assigned by the Document Store whenever the object is created. This identifier is local to the Document Store.

An object can be also identified by means of a path, which is the corresponding sequence of UPIs of the Groups within the hierarchical structure established in the Document Store.

A.5.7 Attributes

An attribute consist of an attribute type and the corresponding values. Some attributes will be defined by International Organizations, and other will be defined by National Administrative Authorities and Private Organizations.

The following attributes have been presently defined:

- Document Store Attributes

  - Members (Information Objects)

- Document Attributes

  - Document Classification (type of content)

  – Creation Date and Time (document creation date not storage date)
  – Document Size
  – Number of references (to the document)

- Group Attributes

  – Membership Criteria
  – Is Temporary
  – Number of Group Members
  – Ordering
  – Size limit (of the group)
  – Type of group (e.g. Cabinets, Folders,...)

- Reference Attributes

  – Is guaranteed (if the referred document is guaranteed)

- Document or Group Attributes

  – Sharing Restrictions (if guaranteed references may be created)
  – Accounting information

- General Attributes (applied to all objects)

  – Unique Permanent Identifier
  – Title
  – Object Type
  – Access Control
  – Parent Identification
  – Statistics

Other document specific attributes and user defined attributes must be added to this list.

A.5.8   Operations

A part from the BIND operation which is used by users to get access to the DS, other operations are defined as follows:

- Document Related Operations

  – Create Document
  – Delete Document
  – Copy Document

- – Move Document

- Group Related Operations

  - – Create Group
  - – Delete Group
  - – Copy Group
  - – Move Group
  - – Update Group
  - – Enumerate Group
  - – find

- Attribute Related Operations

  - – Read Attribute
  - – Modify Attribute

- Document Content Related Operations

  - – Store Document Content (replace the document content)
  - – Retrieve Document Content

- Reference Related Operations

  - – Create Internal Reference
  - – Create External Reference
  - – Register External Reference
  - – Delete Reference
  - – Move Reference
  - – Copy Reference
  - – Replace Reference
  - – Read Reference

The subsetting of operations by defining classes of service is under study and three different classes are now defined: Basic Class, Predefined Group Class and Group Creation Class.

In the version of the standard reviewed in this appendix only the Access Protocol is defined and further communication between DFR Servers is not defined.

Bibliography

[Ami89]   Computer Based Group Communication – The Amigo Activity Model., 1989. Amigo Advanced Group, Redbook.

[BO87]   Steve Benford and Julian Onions. Pilot Distribution Lists - Agents and Directories. In *Proceedings of the IFIP 6.5 International Working Conference on MHS*. North Holland, April 1987.

[BOBW88]   Steve Benford, Julian Onions, Manfred Bogen, and Bernd Wagner. The Implementation of Amigo Distribution Lists. In *Proceedings of the EUTECO Conference*. North Holland, April 1988.

[Bry88]   Bill Bryant. *Designing an Authentication System: a Dialogue in Four Scenes*. Project Athena, MIT, Cambridge, MA, USA, February 1988.

[BW88]   Manfred Bogen and Karl-Heinz Weiss. Group co-ordination in a distributed environment. In R. Speth, editor, *Research into Networks and Distributed Applications*. North Holland, 1988.

[CCI86a]   CCITT. *X.400-Series Implementor's Guide, Version 5*, October 1986.

[CCI86b]   ISO/CCITT Directory Convergence Documents (X.ds) #1-#8, April 1986.

[CCI86c]   Remote operations: Model, notation and service definition, October 1986.

[CCI88a]   Information Processing Systems - Open Systems Interconnection - The Directory - Models, 1988.

[CCI88b]   Information Processing Systems - Open Systems Interconnection - The Directory - Selected Attribute Types, 1988.

[CCI88c]   Information Processing Systems - Open Systems Interconnection - The Directory - Selected Object Classes, 1988.

[CCI88d]   Information Processing Systems - Open Systems Interconnection - The Directory - Abstract Service Definition, 1988.

[CCI88e]   Information Processing Systems - Open Systems Interconnection - The Directory - Procedures for Distributed Operation, 1988.

[CL87]   Shi-Kuo Chang and L. Leung. A knowledge-based message management system. *ACM Transactions on Office Information Systems*, 5(3):213–236, 1987.

[Cro82]   David H. Crocker. Standard for the Format of ARPA Internet Text Messages. Request for Comments 822, DDN Network Information Center, SRI International, August 1982.

[DPB88]   T. Danielsen and U. Pankoke-Babatz. The AMIGO Activity Model. In R. Speth, editor, *Research into Networks and Distributed Applications*. North Holland, 1988.

[DS87]   Draft Recommendations for international public directory service X.500, November 1987. International Telegraph and Telephone Consultative Committee.

[EAN87]   University of British Columbia. *The EAN Distributed Message System Administrator's Guide for BSD Unix, Version 2.0*, 1987.

[ECM87]   ECMA TR-32: Directory Access and Service Protocol, 1987.

[EK88]   Helmut Eirund and Klaus Kreplin. Knowledge based document classification supporting integrated document handling. In Robert B. Allen, editor, *Conference on Office Information Systems*, pages 189–196. ACM SIGOIS and IEEECS TC-OA, March 1988.

[HR88]   Udo Hahn and Ulrich Reimer. Automatic generation of hypertext knowledge bases. In Robert B. Allen, editor, *Conference on Office Information Systems*, pages 182–188. ACM SIGOIS and IEEECS TC-OA, March 1988.

[HT85]   Starr Roxanne Hiltz and Murray Turoff. Structuring computer-mediated communication systems to avoid information overload. *Communications of the ACM*, 28(7):680–689, 1985.

[IFI83]   Naming, addressing and directory service for message handling systems, February 1983.

[Ing85]   *An Introduction to Ingres*. California, January 1985.

[ISO86]   Information processing - open systems interconnection: Specification of abstract syntax notation one (asn.1), May 1986.

[ISO87a]   ISO DIS 8505/4: Message Handling: Abstract Service Definition Conventions, 1987.

[ISO87b]   ISO DIS 8824/5: Abstract Syntax Notation one (ASN.1), 1987.

[ISO87c]   ISO/CCITT Directory System parts 1-8, 1987.

[ISO87d]   ISO/CCITT Directory System, X.ds2 / ISO part 2, Information
            Framework, 1987.

[ISO87e]   ISO/CCITT Directory System, X.ds4 / ISO part 6, Selected
            Attribute Types, 1987.

[ISO87f]   ISO/CCITT Directory System, X.ds6 / ISO part 7, Selected
            Object Classes, 1987.

[ISO88]   Information processing - text communication - distributed office
            applications model, 1988.

[JO87]   Phil Cockcroft Julian Onions, Steve Kille. The Component
            Parts of the PP Mail System. *Internal report*, January 1987.

[Joh78]   Stephen C. Johnson. *YACC: Yet Another Compiler-Compiler*,
            July 1978.

[Mal85]   T.W. Malone. Designing Organizational Interfaces. *Communi-
            cations of the ACM*, 1985.

[MGL+87]   Thomas W. Malone, Kenneth R. Grant, Kum-Yew Lai, Ra-
            mana Rao, and David Rosenblitt. Semistructured messages
            are surprisingly useful for computer supported coordination.
            ACM Transactions on Office Information Systems, 5(2):115–
            131, 1987.

[MGT+87]   Thomas W. Malone, Kenneth R. Grant, Franklyn A. Turbak,
            Stephen A. Brobst, and Michael D. Cohen. Intelligent informa-
            tion sharing systems. *Communications of the ACM*, 30(5):390–
            402, 1987.

[MHS84a]   Message Handling Systems: Interpersonal Messaging User
            Agent. Malaga-Torremolinos, October 1984. International Tele-
            graph and Telephone Consultative Committee (Recommenda-
            tion X.420).

[MHS84b]   Message Handling Systems: Message Transfer Layer. Malaga-
            Torremolinos, October 1984. International Telegraph and Tele-
            phone Consultative Committee (Recommendation X.411).

[MHS84c]   Message Handling Systems: System Model-Service Elements.
            Malaga-Torremolinos, October 1984. International Telegraph
            and Telephone Consultative Committee (Recommendation
            X.400).

[MHS87a]   Draft Recomendations on Message Handling Systems, X.400,
            November 1987. International Telegraph and Telephone Con-
            sultative Committee.

[MHS87b] Message Handling Systems: Message Store, November 1987. International Telegraph and Telephone Consultative Committee (Recommendation X.413).

[MNSS87] S.P. Miller, B.C. Neuman, J.I. Schiller, and J.H Saltzer. Kerebos authentication and authorisation system. *Project Athena Technical Plan (part E.2.1)*, December 1987.

[MTR87] Dwight E. Cass Marshall T. Rose. *The ISO Development Environment at NRTC: User's Manual.* Northrop Research and Technology Center, Palos Verdes Peninsula, California, October 1987.

[OD81] Derek C. Oppen and Yogen K. Dalal. *The Clearinghouse: A Decentralised Agent for Locating Named Objects in a Distributed Environment.* Xerox Office Products Division, Palo Alto, USA, July 1981.

[PS87] Wolfgang Prinz and Rolf Speth. Group Communication and Related Aspects in Office Automation. In *Proceedings, IFIP WG6.5 Conference*. North Holland, August 1987.

[TRG⁺82] Dennis Tsichritzis, Fausto A. Rabitti, Simon Gibbs, Oscar Nierstrasz, and John Hogg. A system for managing structured messages. *IEEE Transactions on Communications*, COM-30(1):66–73, 1982.

[WF86] T. Winograd and F. Flores. *Understanding Computers and Cognition: A New Foundation for Design.* Ablex, Norwood, N.J., 1986.

[Wil87] Paul A. Wilson. Key Research in Computer Supported Cooperative Work, 1987.

Index